Queen Wolf

Robert Franklin Leslie was born in Texas in 1911, of Scottish and Cherokee ancestry. Since boyhood he has explored the wild country of western Canada, Mexico and the USA, often living for long periods of time in remote areas and in close contact with all kinds of wild animals.

He received his BA degree from Santa Barbara State College and his MA degree in botanical ecology from the University of Southern California. As an ecologist and photographer he has lectured widely and conducted photographic tours throughout south-western Indian reservations in the USA, in Mexico, Canada and Europe. As an amateur archaeologist, he has amassed a large collection of Indian relics and lore.

His earlier book *The Bears and I* (also available in Pan) has been translated into five languages and was fictionalized to become a successful Disney film production.

Robert Franklin Leslie lives with his wife in Sherman Oaks, California.

Previously published by
Robert Franklin Leslie in Pan Books

The Bears and I

Queen Wolf

The true story of a friendship between man and wolf

Robert Franklin Leslie

Pan Books in association with **Macmillan London**

First British edition published 1975 by Macmillan London Ltd
This edition published 1977 by Pan Books Ltd,
Cavaye Place, London SW10 9PG,
in association with Macmillan London Ltd

ISBN 0 330 25084 1
Printed and bound in Great Britain by
Richard Clay (The Chaucer Press) Ltd, Bungay, Suffolk

To
my good friend
Gregory Tah-Kloma,
wolf-man of the Kitiwanga

Contents

Author's note 7
1 Náhani of Nakinilerak 11
2 To know a wolf 21
3 An awesome journey 33
4 Trapper-Dan 44
5 The Náhani mystique 53
6 A fragile image 65
7 The danger route 75
8 Eyes in the shadows 85
9 The wolves come home 93
10 The way of the northern wild 103
11 The monarch of Denetiah 111
12 A song of the Kitiwanga 120
13 A wolf betrayed 129
14 Reconciliation 138
15 The hunters 148
16 Silence 154

Author's note

Late one afternoon in the summer of 1970, a young Indian beached his canoe near my Babine Lake camp site in the backwoods of British Columbia. Clad only in shorts, he was tall and muscular, and wore his hair shoulder length. The young man introduced himself as Gregory Tah-Kloma, and told me he was a Chimmesyan of the Tsimshian band. That evening Greg sat by my campfire and grilled salmon fillets for both of us.

During the weeks that followed, Greg and I became staunch friends. We canoed, hiked, prospected, and camped as a team. I learned that he had worked in various mills and mines to pay his way through college. His hands still bore calluses from the work. He was now a graduate student in mineralogy, and spent his summers at placer gold deposits along drainage systems footing British Columbia watersheds. We were both on the way to prospect Babine tributaries when we met.

Night after night, until the black frost of October drove us towards civilization, we sat by the campfire and talked. Gradually Greg told me the remarkable true story of his devotion to a threatened pack of timber wolves, a story that included his search to relocate the amazing female wolf-pack leader, known as Náhani, whose unusual company he had first enjoyed in the summer of 1964. His compelling drive to find the wolf and her pack before trappers and bounty hunters could destroy them reached unique proportions. His fascination for the wolf often took him to the brink of disaster.

I asked Greg's permission to write down his story, and he agreed. He had kept a log in which he listed events in chronological order, and a diary in which he entered his personal feelings and reactions. He allowed me to draw freely on both.

In order to protect the privacy of living individuals and to protect Náhani – who is still very much alive – certain place names and locations have been changed, and various encounters between humans have been slightly altered. However, none of the facts of Gregory Tah-Kloma's adventures with Náhani and her wolf pack have been changed. They are as he told them to me.

Robert Franklin Leslie
March, 1974

There is a pleasure in the pathless woods.
There is rapture in the lonely shore,
There is society where none intrudes ...
I love not man less, but nature more.

George Gordon, Lord Byron

1 Náhani of Nakinilerak

Along the southern belly of British Columbia's north-central wilderness stretches the lake country. Narrow troughs of water up to 130 miles long wind through the canyon lands and resemble wide, gentian-blue belts – their names are Babine, Takla, Tchentlo, Trembleur, Stuart, Nakinilerak, and fifty more.

The longest of these southern 'belly' lakes, Babine and Takla, lie in trenches between the Babine Range to the west and the Hogem Range to the east. A knify Bait Range separates these two lakes. Through a broad glacial canal at the Bait's southern terminus, a series of five smaller lakes, the basin tarns, drain through short creeks that feed Takla.

An ancient Indian trade trail through the Babine-Takla region connects several primitive mountain settlements, often fifty miles apart. The route skirts the upper beach of Friday Lake, northernmost of the five basin tarns. A narrow flume drains Friday into Nakinilerak Lake. Of a winter the trade trail serves aged Carrier Indian trappers who follow axe blazes on the hemlock trunks high above seasonal snow line. Of a summer, possibly half a dozen die-hard sourdough prospectors may dream their way along this obsolete footpath, but no agency maintains or supervises the uninhabited route.

Seven miles south of the old trade trail, a short morning's hike by game runs, lies Nakinilerak Lake, a wilderness gem five miles long, half a mile wide.

In a clump of Sitka spruce and quaking aspens, Gregory Tah-Kloma's camp site straddled a breezy, bug-free peninsula near the lake's intake flume. The year was 1964. About two months remained before a late September or early October snowstorm would hurl him back over the archaic trade route to the totem-pole settlement of Hazelton, where he had left his station wagon with a friend. Prospectors cursed that sixty-mile trek between Friday Lake and Hazelton as a backbreaker, full of deadfalls, winddowns, devil's-claw, icy fords, and landslides. But Greg wasn't worried. His pack would be lighter because he would hide his tools at the 'diggin's'; his food supply would be exhausted; he would throw away his dirty clothes. Much of the route would skim downhill, paralleling the right bank of the Suskwa River.

During the first ten days of July, Greg had panned the stream bed between Friday Lake and his camp site. Thousands of years ago receding Ice Age glaciers had deposited pockets of placer gold nuggets the size of pinheads – and smaller – along bedrock riffles beneath everything from a two-foot overburden of glacial mica up to mountainous moraines.

One morning shortly after breakfast Greg sat rocking back and forth on a driftwood log near his campfire. He liked to finger the two pounds of 'dust' he had accumulated in a canvas bag – a bonanza to supplement his winter salary at the refinery near Prince George. If gold came in any other colour, he reflected, nobody would prize the metal half as much. Chimmesyans say, 'Gold is sunshine stored in a rock'.

As he zippered his precious loot into a rucksack side pocket, he noticed a man trudging up the beach from Nakinilerak's *southern* end.

From a distance the stranger appeared middle-aged. Probably a Carrier, Greg surmised, by the way the man stooped under a tumpline basket tote. Plainly bushed after his long cross-country journey, he leaned heavily upon an alpenstock every four or five steps. With a carbine balanced and clutched at the breech, the man's left arm swung like the shank of a pendulum.

Greg tried to imagine what the Indian had been doing

afoot in that stony wilderness south of Nakinilerak Lake – maybe he, too, was a prospector, maybe a bounty hunter.

'Good morning, sir!' the stranger said as he shuffled up to the smoky campfire. His buck-toothed smile reminded Greg of cartoons depicting friendly beavers. 'My name is Eugene Charley. You have been here long?'

Before giving Greg a chance to answer, he quickly explained that he had been visiting relatives and hunting wolves on lower Takla Lake. He belonged to the Carrier band. (Because of French language influence, Canadian Indians generally say 'band' instead of 'tribe'.)

'I came before the Moon-of-Walking-Thunder,' Greg said. To an Indian that meant early July, because the full moon occurred on the twenty-fourth in 1964. 'I'm Chimmesyan – part Haida, part Tsimshian.'

'You must be gulch-happy. What about *Náhani*? Have you seen her?' As Eugene Charley pronounced the name, he raised his upper lip like a nickering mule. He lowered the lever of his Winchester to check the chamber. The sun shot a brassy glint off a breeched cartridge rim.

Greg urged the man to sit down on the log and remove his heavy pack, the weight of which he bore by the tumpline strap across his deeply grooved forehead. He grinned when Greg offered him a cup of coffee and a pipe stoked with Indian House tobacco.

'Who is Ná*ha*ni?' Greg asked.

Charley spat into the fire. 'You say *Ná*hani,' he said. 'Accent on the *Ná*.'

'I don't give a rusty damn how you say it. Who the devil is *Ná*hani?'

'The great silver she-wolf. Queen bitch of the deadliest wolf pack in Canada. Is this Nakinilerak or Friday?'

'Nakinilerak.'

'They den somewhere near here. I've studied them for a year or more, ever since a sweet price was put on Náhani's head. Those wolves are hunting somewhere south of here. I wish to hell I knew where. When they come back, they'll gnaw your Chimmesyan bones. Nakinilerak is where they winter.' With a speculative squint he probed Greg's expression for reaction.

'Why should anybody be afraid of wolves?'

'Are you armed?'

Perhaps Eugene Charley suspected a rich 'poke' of gold. While he smoked, his glance kept shutting between Greg's gold pan and trench shovel.

'I'm prepared to defend myself,' Greg said without admitting that he carried no firearms. He considered everyone trustworthy until proven otherwise; but this Eugene Charley somehow seemed to speak from two faces. 'Tell me more about your Náhani, whose name you pronounced with such reverence.'

'*Náhani* means "one who shines". Carriers call her Silver Skin. Colour, you know. She's too gutty for a timber wolf – and too damned big. Eight, maybe ten years old. She leads twenty, maybe thirty killers. Who knows? Nobody ever gets a shot at that pack. She can smell a gun a mile away. Livestock killed, traps emptied, and now lately *people* have disappeared. When they raise the bounty enough, I'll bring her down. You'll see.'

Greg concealed his relief when Eugene Charley declined an invitation to rest the day and night. He was headed for Pendleton Bay on Babine Lake. Carriers had to hurry, otherwise the lumber mill would hire Tsimshians to peavey summer-felled logs into the dog-chain lifts. Instead of following the trade trail between Takla and Babine lakes, Charley was short-cutting the route through the brush in order to save time and miles 'and maybe bring in a skin'.

'I'll guarantee you a horrible death if you stay here,' he said as the two men shook hands. 'Náhani's phantom *renégats* will eat you alive!'

'Weasel words!' Greg said aloud. To himself he thought: small-bore talk from a Carrier with a forked tongue, a bounty hunter who builds bad reputations around predators in order to get local authorities to hike the rewards.

*

Nevertheless, Eugene Charley's brief visit set in motion an exciting new trend of thought. Greg's past experience with wolves attested that Náhani and her 'phantom *renégats*', if they existed other than in rum-soaked Carrier imaginations, weren't as dangerous as Charley claimed – unless, of course they were indeed demented renegades, possibly an entire pack infected with sylvatic rabies, blindly revenging themselves against man. He remembered that a leading Canadian newspaper had for years published an offer: a substantial cash payment for any documented record of a wolf having attacked a human being without provocation. No one ever claimed the money. That thought was consoling on a dark and rainy night – yet rabies fell consummately into the category of provocation.

Greg was alone, unarmed. An organized *familistère* (family pack) of timber wolves comprised the most audacious legion of carnivores on Earth. Greg's concern grew as he contemplated Charley's weasel words. But why should he worry more than at any other time? Wasn't danger the inevitable undertone in deep Canadian bush – always?

On the third night following Eugene Charley's visit, Greg squatted near the campfire's orange plumes in order to inhale the fragrance of sweet balsam and cascara bark. An hour after moonset he had watched Ursa Major cross the northern zenith. Not the slightest breeze stirred the forest. For some intuitive reason – certainly not because of any sound or movement – he looked up into the wall of darkness on the other side of the campfire. Barely within the outer perimeter of light stood the largest, most patrician wolf he had ever seen. From Charley's description, it had to be Náhani. Her coat glistened like burnished silver. With intense facial expression she stared back for several moments, then retreated into the dark aisles of spruce and aspen – disappeared as silently as an owl.

Despite certain fears and momentary shock, Greg recognized that the wolf had held her head and bushy tail high – non-aggressive positions. Erectiles of rumps and withers had remained flat, black muzzle and chaps closed, no chomp of jaws. She seemed to be alone although this would be highly unlikely.

Greg retired to his sleeping bag. At midnight a wolf chorus began a series of songs at lakeside. He slipped out and hurried across the tufted sward towards the beach in hopes of glimpsing the pack. But the wolves heard his approach and dashed into dense forest. At greylight they sang again, this time from a headland palisade a mile east of the lake. Náhani and her troop had apparently resumed residence on that hillside. Greg hoped his presence wouldn't interfere with their privacy.

From sunrise until dark the next day the wolves studied every move the young man made. Without so much as snapping a twig or stepping on a dry leaf, they crept like ghosts through the timber, peered from behind shrubs and boulders, inhabited the shadows as shadows themselves. Greg felt their piercing gaze, saw the occasional blur of a movement, but was unable to locate one flesh-and-blood wolf. So he squandered the day, crouching on the beach or straddling the camp log – in plain sight at all times. When he considered it necessary to move about, he did so on all fours, a Chimmesyan technique often employed to make friends with predator mammals. When he tried to mimic their sporadic calls, they replied with chomps and snarls – not hostile, but not friendly. His harmonica tunes and deep-chested chants froze them in rigid, listening silence; but they manifested no actual interest in manmade music.

That evening while there remained two hours of half-moon, Greg decided upon a rash course. Beyond one mile, the beach along the lake's eastern shoreline narrowed and disappeared into rocky, bushy impasses; but since the wolves favoured the hill and forest on that side, he hoped they would allow him to get closer to them in the owl-light if he walked there and ignored them. A feeling of wild ecstasy gripped him as he stepped through the moving moon shadows towards the lake. He had walked less than half a mile when a speeding body crashed recklessly through salal and forest debris on the hillside forty feet above the beach. From size and colour there could be no mistake. She was the big wolf he had seen the night before. Although she allowed him only vapourlike glimpses as she flashed between covering shrubs, she was the most spectacular animal he had ever seen. In spruce-filtered moonlight her coat shone like sterling silver.

At last Greg turned to retrace his steps towards camp. The wolf also dashed in that direction, a little closer perhaps but as yet unwilling to risk any gesture that might be interpreted as a friendly or trusting overture. On the other hand, the young man detected no hint of enmity. He saw many of his own tracks in the damp sand half obliterated by wolf prints; yet he had been unaware of being closely followed down the beach.

The lake was a chilled sapphire. A slip of breeze rocked the forest shadows and waterfowl in cradlelike motion. The closer Greg drew to camp the more conscious he became of large converging masses in the boscage – huge ghosts that made delicate swishing sounds as they moved through duff of dead leaves, dense vine maple, and dry shrub willow. No wonder respectful Carriers venerated the pack as 'phantoms'!

If an attack had been planned, Greg assumed it would have materialized before he returned to camp. But the wolves simply vanished into the hillside forest. Back in his kitchen area shortly before moondown, he stirred up the smouldering embers in the fire pit, stacked a smokeless teepee of squaw wood and spruce cones, then crossed his legs on the spongy sward, leaned back against the old driftwood spruce log, and smoked his pipe. Eugene Charley had spoken with a forked tongue.

On the following morning he sensed a change. From the big hillside east of the lake, the wolves came down – no closer to camp, but they made no further effort to hide, even when they climbed to a rocky spur to sleep in the sun. An atmosphere of relaxation seemed to surround the pack, but Greg decided not to continue his placering. Veteran woodsman that he was, he still felt concern in the close presence of possibly two dozen timber wolves.

From then on, the wolves ignored him by day. They appeared nonchalant as long as he went about camp on all fours; but the moment he resumed the upright position, they assumed a defensive attitude. Somehow they expected him to walk *upright* after nightfall, because that was the way they were used to seeing him on the beach. Older pairs began to relax on the meadow between his camp and the creek, while yearlings and younger wolves ventured to romp up and down the beach or to swim in the lake. Most of the pack swam every day.

Four days later, Náhani sat for more than an hour on a sparsely covered ridge of the hillside. It was the first time Greg had seen her in daylight – even at 300 yards. He observed her through binoculars.

For a week he continued to walk down the east beach at night, and on each occasion six to twelve wolves crept silently through the forest understorey adjacent to the strand. At length Náhani and four adult males walked to within fifteen yards when he sat down a mile from camp and tossed pebbles into the lake. To Greg, they seemed to be probing for a possibility of trust.

Occasionally snorting, whining, snarling, they followed him back until he crossed the flume and returned to his camp site. Footprints here and there revealed cabalistic inspections of his equipment during his absence. Nothing had been touched.

Given that he half expected her on the seventh night, Greg tingled and shivered from head to foot – his teeth chattered, his hands shook – when the long-legged queen wolf stepped slowly out of the darkness and strode deliberately into the circle of campfire light. She sniffed the sweat-laden air, walked to the opposite end of the log, faced the crackling flames, and haunched. Small campfires have always fascinated wolves, and outdoorsmen have often misinterpreted lupine curiosity for aggression. Except for deeply rhythmic breathing, she sat as motionless as a mounted museum specimen.

There is no doubt that something clicked between that wolf and Gregory Tah-Kloma when their eyes finally met. He wanted to believe objectively, without any aura of mystery, that two beings sensed an attraction, one for the other, and at that moment an irresistible friendship began. As an unmated female – as monarch of the pack – Náhani, he felt, was an entity apart from her kind, aloof, alone, and lonely.

Each time their eyes met, further understanding seemed to grow: a bond, shaky and untenable at first, began to bridge the communication chasm between man and wolf. For all her arrogance – no doubt a vital necessity for the queen of a realm – she seemed unable to conceal from Greg an intelligent animal's need for genuine affection. Perhaps Greg was also unable to conceal the same thing from her.

He made no move to get closer to her that night. After an hour of silent 'communication', she stood up mechanically and without turning her head walked slowly towards the hill east of the lake. On all fours Greg followed at her heels until she leaped across the creek and disappeared in the understoreyed forest.

Between moonrise and greylight the next evening, Greg and Náhani sat by the log – almost within touching distance – and listened to the quick of wide-awake wilderness: a bull moose's trumpet, a great grey owl's low-keyed flute, the reedy babel of geese feuding all night on the lake. Their gaze often met as they followed 'shooting star' trajectories behind thin veils of aurora borealis. They watched night itself roll before the amber flood of foredawn, which in latitude 55 occurred at 4:00 a.m. in late August.

There is no calm like summer dawn in the lake country of British Columbia.

Silently Náhani leaped from camp when a team member began an amberlight song. Greg wriggled into his sleeping bag and slept until noon in firm belief that the seeds of trust had sprouted.

2 To know a wolf

From the moment Náhani bounced into camp on the third consecutive night at about eleven o'clock, the big silver female's mood of jubilation found expression in tail wagging, barking, and wide-eyed attempts to communicate a mysterious enthusiasm. Instead of sitting down by the log next to Greg, she trotted around the campfire, head and tail high, muzzle open between utterances, toes spread wide apart. Greg thought her elation strange until it occurred to him that her enthusiasm was not intended for *him*.

Eight dark-pelaged males, each about two stones lighter than Náhani, paced restlessly back and forth over the meadow between Greg's camp and the lake's intake. They gruffed, yelped, and whined like chained hounds, impatiently pleading with Náhani for something Greg could not have understood at that time. Unexpectedly, she placed her front paws on the log and issued a single high-pitched command. Silently, intently obedient, the eight males bounded single file up the game trail towards Friday Lake.

Náhani haunched within eighteen inches of where Greg sat smoking his pipe next to the log. She remained until the males returned an hour before sunrise, at which time she bounded up the trail with about a dozen of her troop.

That afternoon when Greg went on all fours to the creek, six adult wolves and three subyearlings sprawled along the sunny lake shore. Every belly was gorged, which seemed to answer the question of the night before. The eight males comprised a hunting squad. They had brought down game somewhere between Nakinilerak and Friday lakes.

Diffident youngsters lying on the beach stared and whined as the young man edged within two log-lengths, but the adults dozed in peace until he crawled to within three yards of where they lay. After a quick identifying glimpse, they ignored his presence among them. Each wolf from his own sentry post had seen Náhani sit with Greg by the log. She had certainly communicated with her own.

Except during the Indian's moonlight hikes along the lake shore, Náhani had not appeared either in camp or among the other wolves before eleven o'clock of an evening. However, at homing time on the afternoon following the night of the hunt, she trotted down the hillside, jumped across the creek, and entered camp for an early visit. From belly dimensions it was plain that she, too, had feasted on a carcass.

In his diary Greg described the manner in which Náhani projected monarchy over her entourage from the moment she stepped on the beach where most of the pack lay in torpor from overeating.

'She must weigh two stones – maybe two and a half – more than her biggest lieutenant. In animal societies, weight, speed, and guts count for most. Add brains and you come up with leadership. Náhani has all four. Every wolf awakes and springs to attention when she shows up. After a quick round of rear-end smelling – to make sure no stranger has joined the pack – she starts brow-beating the other females, to make sure they remember who is boss. When one sassy old female with sagging teats, pot-belly, and arched tail failed to roll over and expose her under parts in submission, Náhani grabbed her by the ruff and threw her on to the rocks. Had the sassy bitch moved or growled, Náhani might have killed her. After that example the males showed readiness to obey: tails between their legs, whines, licks, and bellyups. I've got a hunch Náhani holds on to her subjects because she personifies everything a wolf respects. She kowtows to no special male. And that means she either lost a mate or never had one. It is plain funny the way she about-faces and showers affection on every wolf in her pack – after she chomps hell out of a few. Lifts her hind leg and pees on rocks and stumps just like a male.'

According to Eugene Charley, Náhani and her phalanx ranged fifty miles south of Nakinilerak to Trembleur Lake and fifty miles north to the Shelagyote Glacier in the Nilkitkwa watershed. It was surprising that Carrier Indians ignored the actual range which covered at least twice that distance. Because Charley reported that his relatives had spoken of little besides the elusive 'phantom pack', it was reasonable to believe the wolves had hunted far south of the so-called 'belly' lakes during Greg's first days at Nakinilerak.

In such an event they could very well have reached the northern rangeland of domestic livestock.

Given that any adult pack member could cover seventy-five miles or more in four hours, Greg assumed that scouts had reconnoitered his camp at times and then 'reported' back to Náhani wherever the pack had holed up along the southern extremities of their runway. He considered it conceivable that Náhani herself had studied his habits long before moving her charges back to the Nakinilerak denning complex. Restless curiosity often outweighs the wolf's quest for security, often leads him farther than necessity will drive him.

One point was clear. The wolves had reoccupied their quarters only after observation of the man camped near what may have been their ancestral whelping dens. This is common practice among northern canids.

When the pack more or less accepted human presence near their hillside, Greg resumed most activities, including placering, on foot rather than on all fours. The first time he walked back to camp from the creek, however, fourteen smoky-brindled forms sprang from a sleeping position on the beach and raced to deep forest safety. Sentries had sounded a subvocal alarm at the unusual daytime occurrence of the man-walk position.

Although the pack quickly tolerated his presence – as long as he kept his movements routine – one old pair of black, heavily mantled individuals always scooted away with parted flews and chomping teeth if he approached except on all fours. Both black wolves limped. When those two slept on the sand or the sward, Greg continued to respect their preference. He discontinued all placer operations when he discovered that the work disquieted the entire pack.

Náhani was first to accept the man-walk position. One evening she trotted into camp after Greg had finished ash baking a sockeye-salmon dinner. On several occasions she had refused offers of *raw* fish but gobbled it baked or fried. She accepted a second fish and took it to the hillside beyond the creek, which meant she was probably succouring an incapacitated associate in a den.

She returned immediately to camp. Haunched by the log, she eyed a number of activities that Greg deliberately performed on foot. Quite suddenly she jumped squarely in front of him and blocked passage. At first her act appeared to protest his upright position; but finally she reared, balanced around for a few seconds on her hind legs; then placed her forepaws on his shoulders. The precipitate shock of her 140 pounds almost bowled him backward into the fire; but more than the physical shock, that huge head twelve inches wide, maws open, two-inch daggerlike fangs so close to his face caught him unprepared. She simply licked his nose, gently shoved him, and backed away. He was to learn that Náhani responded much more swiftly than any animal he had ever known. Her headlong acts visibly revealed a complex of deep-seated emotions that often surfaced without prior announcement.

In addition to her sensitivity, Náhani's every physical characteristic struck Greg as beautiful and satisfying: her extraordinary silver coat; her bicycle-pedalling gait that broke into an effortless canter like that of a thoroughbred horse; her wide-set, greenish-amber eyes; the muscular rhythms in her powerful shoulders, neck, and legs when man and wolf raced down the beach; the heavy, foxlike tail, two feet long and tipped with black, whose position and hackles expressed her every whim and mood. To Greg, of course, her comeliest attribute was the new routine she often employed – that of placing her forepaws on his shoulders and licking his face.

During his first three weeks of association with the wolf, Greg made no move to soft hand Náhani, not even when she came to lean against him near the campfire after long runs down the beach. He had known people with pet wolves who often claimed that their animals enjoyed caresses from the human hand, but he considered it unmeet to touch any wild creature until positive the gesture would not be misinterpreted.

Some days before he began petting Náhani, she developed a nuisance habit. She took his arms, hands, or feet into her mouth and chewed. The action probably demonstrated affection, but it may have been to deposit her own salivary scent, thus proclaiming dominion or ownership even to the exclusion of friendly overtures from other wolves. As pack matriarch, she never engaged in frivolous play. When moulting pelage caused her skin to itch, Greg seized the opportunity to run his fingers through her coat – to curry loose hair by the hour – to rub her head, back, and belly. Under those circumstances she learned that *unchewed* hands and arms performed better services, because her friend refused to comb and scratch when she insisted upon chewing human extremities with teeth and jaws designed to fell moose and elk.

Repeatedly six to ten subordinates stood fifteen lengths from the log and fox-eyed the combing and rubbing. They had seen her place paws on Greg's shoulders and lick his face. They had never been far away when the two walked or raced along the lake shore or sat for hours in close physical contact near the campfire. Yet, on the surface at least, Náhani was the only wolf in the pack interested in Greg's company. At first it annoyed him, because he had tried to make friends with several more approachable wolves. Two small twin females wiggled and wagged their tails whenever they passed by his camp. Then he recognized that Náhani herself, through the scent of her urine, saliva, caudal-gland secretion, and paw sweat had forbidden the others to have anything to do with him. Jealousy and suspicion pervaded the colony, but the dominant trait of every wolf was loyalty to Náhani's wishes.

Apparently satisfied that combing and rubbing would do their queen no harm, the other wolves retired to the open beach to chant solos or sing in chorus. Greg was unable to associate their songs with any form of lupine communication other than an emotional expression of the moment. They never bayed at the moon. Baying appeared to be a kind of family excitement and nothing more. The true wolf song was in no way related to hunting, mating, calling, or 'mooning'.

Greg was never sure that Náhani sang. In his presence she merely whined when midnight serenaders reached certain pitches of eloquent harmony. And eloquent they were – to Greg the wolf songs seemed the purest animal music in all nature.

At times the colony was an anthill of laudable activity. Náhani coerced all wolves without young to feed, protect, and train other pack members' cubs. Apparently each adult did regular sentry duty. Cleanup committees periodically removed bones, fur, and ordure from the dens or helped with excavation in the event of cave-ins or retunnelling. The matriarch assigned special guards and helpers at the den of any sick or disabled wolf. Her chieftainship was perhaps most conspicuous each time she assembled her hunting squad and either led them or sent them towards the scent of game. Her ability shone when she drilled and timed each member's duties for offence, defence, or escape. After each kill, the subalterns led the non-hunters to the victim's carcass. When they had fed, they brought meat and bones to bury near the dens for later consumption.

Machinelike only when they followed Náhani, the wolves demonstrated wide individual differences in intelligence and disposition, especially so when they were 'off duty'. Notwithstanding an inflexibly observed chain of command, they evidenced tender affection and loyalty towards one another throughout the several generations represented in the pack. There was neither petulance nor fighting among the several families. In play they adhered strictly to the 'peck rights' principle.

Late one afternoon Náhani and Greg went berrying along the stream between Friday and Nakinilerak lakes. When they returned, twelve members of the pack were on their way to escort them back to camp. They were obviously in a state of agitation. They led the way across the meadow and towards the hillside. A solitary voice drifted down – a tremulous, sorrowful wail. Greg ran after Náhani as she jumped across the creek and loped up the steep, spruce-crowded slope east of the lake. Gasping for breath, he finally reached the granite merlons at the skyline crest, barely aware of having passed through the pack's secret denning complex. While he scrambled along the rocky ridge above timber line, Náhani and most of her legion crossed the summit's east flank. When Greg arrived, the wolves were baying in a state of high-pitched rage, lunging and slashing at a huge grizzly bear – in Chimmesyan dialect, Ozilenka.

The 800-pound giant had stalked and killed a truant wolf cub. The mother's moan had signalled the crisis. Erect on his hind legs, the bear stood nine feet tall as he pivoted and swung at fifteen wolves who had just seen their sanctuary outraged by their only natural enemy.

Like a Haida chieftain, Náhani commanded each organized charge. Twice the bear's deft right hook intercepted her lunge, but she allowed herself to fall limp into a heath cover of button willow. The bear, of course, recognized her as the leader. He knew that her offer of herself as the target spearhead for each attack was to distract him long enough for pincer wolves to move in on both sides. The grizzly could count on certain victory if he quickly dispatched the leader. Although outwitted in strategy, the bear was fast enough and strong enough to extricate himself from each of Náhani's manoeuvres.

Her guttural baying goaded the pack to more furious attacks at the grizzly's loins and shoulders. Instinctively, the wolves respected the most powerful swat in the animal kingdom; therefore, they never attacked a grizzly head-on. His tough skin and dense pelage easily resisted their vice-like jaws and dagger fangs.

The wolves lost a measure of fury when the bear hammered two of their number to the ground – one with a broken neck, one with a crushed skull. Finally the panting pack bayed as one voice and paced a wide circle around their unruffled foe. It was a breather in preparation for Náhani's next move.

Expecting the wolves to back him up at this point, Greg seized two baseball-sized rocks and bounced them off the bear. At the same time he yelled a tribal war whoop that frightened not only the bear but the wolves. He had already estimated the distance to the last timber-line spruce in case the bear charged him. He depended upon the ancient precept: adult grizzlies don't climb trees.

The bear dropped to all fours, locked his jaws around the dead cub, and lumbered down the hillside with his prey. Every wolf slashed at the retreating grizzly's hind quarters. The noisy mêlée disappeared among moorland boulders too distant for Greg to follow.

On the way back to camp, he explored dugouts beneath deadfalls and rock overhangs. Subyearlings and the two small females snarled when he snooped into their dens.

*

At nightfall Náhani dragged into camp and dropped at Greg's feet. For a time she lay panting and whining. He lifted her head to his lap and, until the Dog Star set, combed dried blood and saliva from her coat. He found no wounds other than bear-claw welts along her sides and neck – welts the size of shrew runs. The campfire light shone like an overglaze on her eyeballs. She was plainly exhausted but continued to rumble with the fire of hatred. Her lieutenants moaned, yipped, and yowled as they war-danced around the camp site.

Instead of singing at greylight, the wolves snarled, chomped, and whined, slogging back and forth between camp and the denning complex as if looking for the three dead members of the pack. Greg recognized at once that they were begging for leadership to pursue that grizzly and resume combat. At dawn, Náhani walked stiffly towards the dens.

After the battle with Ozilenka, the pack regarded Greg with less suspicion; but irrespective of the degree to which each individual now trusted him, no wolf dared violate what the queen-leader had pre-empted. His friendship with Náhani flourished exclusively on the wild terms of its inception – uncompromising independence. Trust was the added ingredient; and in order to keep that trust, Greg gave up further attempts to cultivate friendly relationships with other wolves. For days following the tragic episode, Náhani drilled her fighting team along the ridge where the grizzly had been. She divided her forces into two squads that practised brutality against one another until Greg wondered if an organized pack would survive the violence.

At no time did he suggest to Náhani any form of obedience, even during her recent moods of hot-blooded ill-humour. In deep humility he honestly admitted his obedience to the wolf. When she chewed his boots, gobbled his food, destroyed his only bath towel, and urinated on every item of his equipment, he was careful to allow no resentment to enter his expression. Day or night, only one bark was needed to bring him to her on the run, to follow her along the shoreline, up the creek to Friday Lake, or over the rocky ridge to sniff for fresh bear tracks. When she craved a salmon dinner – and ordered it by bringing fish-bones to his camp – he fished and cooked for her. Once he had memorized her physical signals, her communication was generally obvious.

In a sense, Greg became one of Náhani's subjects. He experienced intense joy obeying her, and felt a primitive bond that defied analysis. When he realized that the days of good weather were slipping by, he stretched those remaining hours with the wolf by cutting his own sleeping time, by sitting with her while she slept in the sun.

*

When any living being is threatened with premature death, the Chimmesyan believes that the individual has stumbled into the shadow of a spear. Thus, when Ozilenka the grizzly bear returned, killed a second cub, maimed an adult wolf, and went away unscathed, Greg foresaw disaster for the pack. To compound the problem, on September 5 an icy blizzard from the Beaufort Sea drenched and frosted central British Columbia. Balsam poplars, maples, aspens, dogwoods, alders, and sumacs changed into spectacular autumn dress even before the storm blew itself out. Suddenly black frost stripped every hardwood, desiccated every herb and forb, sealed off the summer ferment in bog and duff. Only the antiseptic smell of resin remained in the rime-chilled air. Storm shadows intensified the conifers' sombre green.

On the roily night of the eighteenth, Náhani's entire pack assembled around the perimeter of Greg's camp, squinted their eyes into the frost-heavy wind, raised their throats, and sang to the grumbling cloud mass overhead. At first Greg interpreted the serenade as a statement of friendship, but better judgement warned him that the song was undoubtedly a medley of command calls. He and Náhani had huddled for hours in the lee of a tarpaulin lean-to and watched snowflakes sizzle on campfire embers. When the great white wolf finally turned her face, Greg saw a faraway expression come into her eyes.

The queen wolf had reached her inevitable decision, her inevitable response to those calls. After one swift lick across Greg's face she sprang to her feet and leaped into the stormy darkness. She yelped for her pack to follow.

'Náhani!' the young Chimmesyan called several times, but of course she had no understanding of what the word meant – even if she could have heard him above the wind, above the vigorous baying of the pack. *Náhani* was man tongue. Greg had never used the name in her presence.

The wolves skirted Friday Lake and headed up the pass into the Nilkitkwa basin. Any Canadian woodsman could have decoded the message on those southing pillars of polar air. The wolves had scented woodland caribou herds migrating to their winter quarters.

By October's Fallen-Leaf-Moon, Náhani and her legion

had not returned. Greg had no more food. He broke camp and shouldered his haversack. The best Chimmesyan medicine men never predict how long an autumn storm will last. And soon the deep snows would bury the only route out of the wilderness.

3 An awesome journey

Greg returned to Nakinilerak Lake under the June Moon-of-the-Moose-Child, 1965, to prospect for gold and to renew association with the wolves. Gold he found, but Náhani and team did not return. Fishers, foxes, otters, and coyotes battled for possession of the abandoned hillside dens, so well constructed. During the autumn and winter moons, Greg questioned Babine, Beaver, Sekane, and Carrier Indians about the Silver Skin. They all assured him that Skeena bandsmen had reported sighting the 'phantom clan' on the Stikine Plateau to the north, where Omineca meltwater fountainheads three mighty rivers. The wolves had acquired this bountiful range during the winter of 1964–1965 either by force or by confederacy with a resident pack. The Stikine country in the north-central British Columbia is the last North American frontier of extensive temperate-zone wilderness remaining in primeval state.

To the Ancient Ones – the aborigines – this isolated region, walled against human trespass on the west by the Coast Ranges and on the east by the towering Rocky Mountain cordillera, was known as the Kitiwanga. To all but a few woodland Skeenas, Tsimshians, and Chimmesyans, the word has lost its historical meaning – 'not for man'.

In the short span of two years, Náhani became a legend among northwoods Indians. For alleged crimes against the Dominion – mushrooming tales disseminated by bounty hunters like Eugene Charley – there was a growing price on her head.

Presenting himself at the spring meeting of the tribal council, Greg asked the sagamores for wisdom in his determination to relocate the wolf and her pack. They shook their heads.

'Náhani,' they responded, 'is bad medicine!'

Greg consulted aerial survey maps. After weighing the hardship, expense, and massive odds against ever finding the silver wolf in a pathless wilderness labelled by its own natives 'not for man', he returned to his job in Prince George. When not at work he constructed specialized backpacking equipment and dreamed of renewing his search.

*

On June 10, 1966, Greg arrived at the auto repair shop of his boyhood friend Rocky Longspear in the Tsimshian village of Hazelton. It was almost twenty months after he had last seen Náhani and her wolf pack. During that time many reported sightings south of the Skeena River and tales of depredations by the 'phantom pack' had drifted into Canadian communities on the tongues of hunters and bush pilots anxious to keep the fantasy alive. At last the great Kitiwanga snows of two consecutive, severe winters had melted, and Greg could now begin his search for the wolf. Rocky garaged Greg's station wagon.

His outfit included the bare essentials of lightweight survival gear and a ninety-day supply of freeze-dried and dehydrated foods; but even after removal of every non-life-sustaining item, the pack weighed four stones – almost sixty pounds.

Two seasoned Tsimshian guides who were lounging around the auto repair shop eagerly offered their services until Greg revealed the object of his trip. To prove he was 'off his rocker', they took him to Trapper-Dan Tall-Totem's cabin on the Kispiox River Road. Dan had trapped along the Skeena River in the Chettleburgh country during the preceding ten winters. When the trapper answered the knock at his door, Greg at once interpreted the apathetic expression and bloodshot eyes as danger signals – here was a combustible man. One of the guides explained the reason for disturbing the trapper.

'You gotta be nuts!' Dan growled with an unmistakable tone of hostility. 'For two seasons your big Náhani and pack destroyed every trapped skin north of Kimolith Creek. They dug and chewed out the sets and run off with 'em. Three strings I lost.'

He revealed a recent meeting with Tlingit and Skeena bandsmen from the Stikine River country in the north. They had experienced identical misfortunes. The trappers' association had upped the stockmen's reward for Náhani's hide; even the lumber companies had contributed, because timber cruisers and fellers refused to work southern plateau leases without armed guards. The man who could bring in Náhani's pelt would receive $800.

'Come Moon-of-Walking-Thunder, Mr Wolf-man,' Dan said as he pointed a shaky finger towards Greg's face, 'you can kiss your playmates goodbye. We all heard that damn yarn about you buddy-buddying with the wolf bitch at Nakinilerak. Every trapper in the Kitiwanga meets at Kimolith Creek in July. We won't sleep till Náhani is dead and skinned.'

'How do you know where she is?' Greg asked.

'We'll get there before you do, wherever she is. There ain't no trails in the Kitiwanga. But we know that country like the palms of our hands. You don't. That's all.' He backed inside and closed the door.

Perhaps a dozen Hazelton Indians besides Greg's friend Rocky Longspear honestly believed that Greg had associated with Náhani and her pack. Rocky had closed his garage during the dreadful winter of 1964–1965 and had gone to work as a mechanic at the Pendleton Bay mill. It was there that he met the talkative Eugene Charley. He had not convinced Charley that Greg had come out alive. News that the 'wolf-man' was a guest of the local mechanic spread throughout the totem-pole-lined village within minutes. Feelings and superstitions concerning Náhani and her pack ran high. Most of the six hundred inhabitants depended on 'the goods of the woods' for a living.

'Get out of town as soon as possible, Greg,' Rocky warned. 'You are best off not to trust Trapper-Dan.'

'I've got to find her now, Rocky,' Greg said emphatically. 'I have to get her farther north if possible. With that reward dangling in front of their noses – and these blockheads gobbling up everything they hear . . .'

*

A wet chinook blew in from the coast. For the next two days Greg sloshed through wind-driven rain. The dim trail led through second-growth forest and spiny undergrowth. When he waded into the sudsy mud at the settlement of Kuldo, all but one of the cedar-planked, matchbox shacks on the mighty Skeena's right bank were deserted. A wrinkled, toothless squaw who called herself Moiso invited him to spend the night in her one-room hovel, which proved far less attractive to Greg than the dripping cottonwood grove where he had planned to tent out.

The roof leaked, the only window had been nailed shut, and a marmite half full of overaged venison liver gurgled with wild onions on a cracked and smoky cast-iron stove in one corner. An open slop jar that hadn't been emptied in a cat's age reeked alongside an unoccupied space on the floor where the old squaw motioned for Greg to unroll his sleeping bag. Despite the rank atmosphere, Moiso's hospitality kept the young man from bolting back into the rainstorm. She explained that Kuldo villagers had packed out to swap winter pelts and smoked salmon for staples at the trading post in Smithers.

'Pelts?' he asked. 'I thought Náhani had destroyed the sets.'

Moiso all but choked on the hunk of liver she had been gumming. Through cloudy green eyes she stared at the nailed-down window. After an adequate burp she whispered, 'Crow talk!'

'You know about Náhani?'

'She came to Kuldo last winter. Witch doctors and sachems speak about the deadly white phantom.'

'Trapper-Dan Tall-Totem said his clan aims to kill her under the Moon-of-Walking-Thunder.'

'So they plan. They may kill her. Trapper-Dan killed plenty renegade wolves before. But the Kitiwanga goes to the end of the Earth. Náhani runs farther and faster than any man ... or any wolf before.'

Bear Lake Carriers, she said, saw Náhani east of the Sicintine Range. A bush pilot reported a large pack led by a silver-white in the Skeena Mountains. Moose poachers in Driftwood Valley swore Náhani's pack stole their kills all winter.

Later that evening Moiso directed Greg to follow the Skeena's right bank until the river flowed around Mount Tommy Jack's northernmost shoulder. It would be smarter to return to Prince George, she declared, and earn the white man's wampum; but if he was determined to commit suicide he should camp at Tatlatui Lake, headwaters of the Finlay River, and wait there for Náhani. Moiso had camped there with her father when she was a girl; she claimed to have seen many white wolves at Tatlatui. According to Moiso, *Tatlatui-meh-Náhani* meant 'home of the shining white wolves'. All silver skins came from and returned to that region.

As soon as greylight opened the trail, Greg gave the old squaw a dollar and started up the wide Skeena River gorge. Until the footpath petered out he made fairly good time despite hungry swarms of deer flies, black gnats, and mosquitoes taking advantage of a sunny day and fresh Chimmesyan blood.

Ten miles beyond Kuldo the teeming Kitiwanga game herds began to appear. Mountain goats and bighorn sheep on Tommy Jack's broad shoulders trotted to the riverbank and ogled when they saw the young Indian on the sunset side of the stream. Beyond the big bend, where the river flowed northwest through a gap in the Sicintine Range, he overtook moose, wapiti, deer, and tagtail coyotes.

Beyond the big bend, the river flowed in a deep, unveering trench for about forty miles, the first half of which demanded four and a half days, twelve to fourteen hours a day, of racking, complicated trail blazing. Because of cliff-like banks, Greg often considered himself lucky to walk fifty yards at water's edge. Then the only way through was to scramble to haggard heights 2,000 feet above timber line. For miles the dense spruce, cedar, and hemlock fostered an understorey of impenetrable shrubbery laced with cat's-claw, devil's-club, and spiny eglantine. The underbrush thinned out only when deadfalls and winddowns overlaid the steep slope. To climb and jump from log to log with the heavy Trapper Nelson pack rack and rucksack was continuously hazardous. Deep sphagnum moss carpeted hidden layers of slippery fungi and liverworts. Fearing a fractured bone, even a serious sprain in that remote jungle, Greg slackened his pace even though he knew he was losing valuable time.

Low-skirted spruce, branching almost to the level of forest debris, further impeded his progress. Despite mosquito clouds, deadfalls, low branches, and steep terrain he felt intense fascination for the savage Kitiwanga with its radiant life forms. And always there was the goal of finding Náhani before the trappers and bounty hunters reached her.

On the latest Dominion aerial quadrant maps Greg marked his progress each day when the tree shadows were shortest and again when they were longest, in order to assess his exact position. During those first days, he also double-checked his position against a precisely declined compass, and triangulated against recognizable landmarks such as Chettleburgh Peak and Mount Tommy Jack. He was obsessed that he might fail to mark off on the map one creek or minor change in direction. Any such mistake might necessitate a devastating climb to timber line – often 3,000 feet – in addition to the tricky mathematics of triangulation in order to re-establish exact position. Actually, as he finally realized, it was all but impossible to get lost as long as he followed the narrowing Skeena River canyon. He might have cost himself an extra twenty-five miles, however, had he failed to leave the big river gorge at Iceberg Creek. The insignificant little waterway, he discovered, joined the Skeena under woven willow and alder overhang, as if intentionally screened from right-bank view. But Moiso had instructed him well.

At the Iceberg confluence the resounding Skeena, fifty yards wide, cackled through a quarter mile of braided rapids – fast, deep, cold, and deadly. No possible ford. A twenty-five-mile hike upstream in order to get beyond the volume that numerous tributaries added meant fifty miles, with the return trip down the left bank. Swimming the Skeena's powerful current with a four-stone pack was out of the question, rapids or no rapids.

An excellent camp site with dry wood and few insects extended above the cobbly beach opposite Iceberg Creek.

He wrote: 'I pitched camp, hooked a ten-pound sockeye, and ate supper. Worried for the next eighteen hours. A yearling bear – a funny little imp – horned in and helped me polish the fishbones. An old female wolverine clobbered the bear out of camp and made it plain I should feed her. The gutty bitch gulped salmon, char, trout, and suckers with equal verve and faster than I could catch them. But I admired her guts.'

He went on to characterize the wolverine as an unremitting pest – greedy, monstrously ugly in full moult, and unafraid to the extent that he considered her dangerous even with a full belly. Too much crisp excitement sparkled in her eyes, and the whole canyon smelled of musk that continuously leaked from overaged, overworked anal glands. She muttered incessantly.

As he sat by the campfire that night and played his harmonica, trying in vain to get that river crossing off his mind, the wolverine darted through camp every few minutes, wheezed, snarled, and hissed with startling shock effect. She came and went through the night but somehow respected the young man's privacy once he retired to the sleeping bag.

*

Getting across the pitching Skeena was one of Greg's major undertakings. There seemed to be no choice but to continue up the right bank. He had walked less than half a mile alongside the frothing stream when he came upon a mass of driftwood logs. A raft! He spent the morning extricating suitable lengths and lashing them together with the nylon rope he carried for animal-proofing his food at night. He cut a living spruce sapling to provide a ten-foot drag-brake pole for slowing the raft in swift water.

After tying his clothing and pack rack to the forward end, Greg eased the raft into the current. It submerged immediately when he got aboard. The logs were water-soaked beneath deceptively dry outer surfaces. So he decided to walk across. Fifteen feet out from shoreline the raft collided with the first whitecaps. The water was ten feet deep. All Greg could do was float downstream and hang on to the hindmost. At the first cataract the raft became wedged between two underwater boulders so abruptly that Greg's breath was knocked away and his chest scraped. He couldn't work himself in front of the heavy contraption to lever it over the rocks for fear of being crushed when the current carried the raft forward. Slippery algal slime on submerged boulders posed another problem.

At length, with the drag pole, he pried the raft over the rocks. The clumsy mass no sooner refloated than the vertical current lodged it against the next impediment. A chest-deep shoal, however, allowed him to stand on the gravelly bottom long enough to rock the sturdy boom of logs across the barrier. The spouting current spun it immediately out of Greg's control. A side eddy beached the raft within twenty feet of his previous night's camp site.

Adding irony to failure, there on the beach stood that abominable wolverine, chattering in what Greg was inclined at that moment to construe as a horselaugh.

His body was too cold to feel the sting from a scraped chest, but he broke out in a rash of itchy welts. He untied the logs. With the rope he strung his food over a limb, then built a bonfire to restore his blood to normal circulation. Once his fingers regained sufficient dexterity, he assembled the fishing gear and stumbled to the foot of the rapid. From deep, slow-purling pools, he snagged four large salmon. One for himself and three for the wolverine. She left after eating the third fish and did not return.

As he was angling in the green-water pools, Greg recognized his navigational mistake. Instead of attempting to cross what he misjudged as shallow rapids, he should have cordelled the logs along the shoreline to quiet water. At that point he could swim behind the raft and push it across the Skeena with only the inconvenience of half freezing.

That night he slept soundly from dark to daylight. Shortly after breakfast he reassembled the raft and swam it to the left bank in less than half an hour, but the current carried him a mile downstream before he could beach. He built a thaw-out-fire, then untied the logs, and proceeded to Iceberg Creek. A thickening buttermilk sky mantled the sun.

With rain imminent, he erected a Chimmesyan lean-to shelter from driftwood timber, chinking the cracks with reeds and rock-weighted sphagnum, and gathered a three-day supply of dry squaw wood. His present camp site was directly opposite the one on the right bank fifty yards across the rapids. Greg realized ruefully that he had made very little headway during the past five days: an average of ten yards a day!

On the fourth morning after the crossing, the sky promised a week of good travelling weather. Impatient to reach Tatlatui Lake after the delay, Greg began the long slog up the obstacle-filled canyon of Iceberg Creek. He had considered the first ninety miles beyond Kuldo among the most difficult terrain he had ever encountered. The route between the Skeena River and the lake was to prove even more exacting. The distance, less than thirty miles, required six and a half suns so dense was the forest, so rough the terrain along the Stikine Plateau's western approaches. (British Columbia Indians measure a summer *sun* as a span of about nineteen hours, from dawn to dusk.)

Pleistocene ice had glaciered the southwestern flanks of the Stikine Plateau. Canyons, broad rolling hills, jagged crests, and soggy undrained meadowlands zigzagged through and around belts of wind-gnarled spruce, jack pine, and stunted hardwoods. Thick copsewood timber was even more treacherous to penetrate than larger-boled hemlock and cedar forests. After the deep Skeena gorge, Greg's eyes were unaccustomed to hundred-mile horizons. He had to overcome a compulsion to climb above timber line every time he found crests that accommodated a view. Nevertheless, at the expense of severe insect bites, he was unable to resist detours into meadows and upland fens where arrays of wildflowers contrasted extravagantly with the monotone of unending forest.

At Tatlatui Lake, Greg installed a base camp in a cove at the upper end. Grayling, squaw fish, and char seized every lure that moved. Bitterns, coots, and loons fluted night and day from reedy inlets. Geese and ducks fought for sedge brakes and never ceased their honking. Moose, elk, goats, and deer jostled through the camp site as if the presence of man posed no threat. Throughout the day the forest resounded not with the music of summer songbirds, but with cries of alarm. On regular night-time feeding circuits, lynx, bears, foxes, coyotes, and cougars glided between camp and the glassy, moonlit countenance of the lake. None of the animals made the slighest effort to hide itself from view.

Wolves were the only missing creatures. Recalling what had happened at Nakinilerak Lake – and old Moiso's prophetic words – Greg simply sat tight, hoping Náhani would come to him. Under the first four suns he hiked along the lake front and climbed the ridges behind the basin. A kind of pre-wolf silence pervaded.

He waited.

Then one morning at sunrise as he was about to light a breakfast fire, he sprang to his feet. The easterly breeze brought to his nostrils the unmistakable smell of coffee and sizzling sow belly. Someone had camped near the outlet flume. With a flood of apprehension, Greg realized that Trapper-Dan and the bounty hunters had arrived.

4 Trapper-Dan

Greg scooped up his binoculars, slipped his belt through the loops of a side-knife scabbard, and began the five-mile trek towards the source of his concern.

He ran along the first three miles of the beach, blaming himself for the number of footprints he had left on his first encirclement of the lake. Trapper-Dan was well known as the smoothest tracker in the province. He would deal harshly with any man who tried to stand between him and the reward for the queen wolf's pelt. But the reward was a side issue as compared to the glory he would receive after bringing in the skin of Náhani.

Trapper-Dan was an opportunist. He clung to no superstitions, had no religious belief, no tribal tradition, no monitoring conscience. He regarded the taking of human life with the same lack of concern he showed for a trapped fur bearer – less, because he could *sell* the skin of a fur bearer. According to Rocky Longspear, everyone, including the RCMP, dreaded an encounter with Trapper-Dan.

A mile from where he could see the campfire smoke that curled into the morning breeze, Greg left the beach and climbed to a ridge commanding a dioramic view. One look through binoculars confirmed his worst fears. There, sipping coffee, smoking, and polishing rifles, sat five men – Trapper-Dan, three white strangers, and the beaver-faced Eugene Charley.

Even more disconcerting was the five-place float plane beached near the lake's outlet flume. How could an airplane land on that lake, Greg wondered, without his having heard it? Perhaps the pilot had cut the engine long before the approach, hoping to surprise the predators if they were in residence. Knowing that wolves spooked at the throb of an engine, pilots with hunter fares generally switched off their power and glided over the treetops to level off on a lake 'dead stick'. A man on each pontoon then paddled ashore.

After a one-minute study of the situation, Greg carefully retraced his steps to his own camp. He packed quickly, then obliterated as many traces as possible of his having camped there. For two miles during his retreat into the range south of Tatlatui Lake, he erased his every footprint. When it was practicable, he walked on stones, gravel, or dry duff.

There was no visible pass through the range, and the perpendicular limestone cliffs along the entire north face prevented further advance. Greg could see a gulch that ran east, but that would take him back within two miles of the Tatlatui outlet. Although he dared not risk an encounter at that point with Trapper-Dan and his allies, the only choice was to drift carefully down the narrow ravine towards Tatlatui Creek. At sunset he walked into the gap where the dendritic Tatlatui-Kitchener drainage system formed the Finlay River. Stepping on solidly packed glacial gravel, he experienced a momentary sense of comfort; but as the gorge deepened, thick shoreline growth often forced him to slosh along up to his knees in water. His progress was discouragingly slow. Because of the rugged terrain, it was impossible for him to climb beyond the timbered chasm to check his surroundings.

After marking off several tributaries on the map, he struggled into a box canyon for the night. Still too near the enemy camp to risk a fire, he ate dried fruit and a candy bar before retiring for a long night's sleep.

The next morning, rose vines and raspberry brakes caused his descent of the Finlay River canyon to become increasingly slow and hazardous. Two days dragged by before he splashed abruptly into the river-like Thutade Creek. At this point he left the Finlay and climbed seven miles to the debris-jammed outlet of one of the most resplendent bodies of water in Canada. Thutade Lake, occupying a deep trench between thunder buttresses in the Omineca chain, is about three miles wide and winds for thirty-seven miles through spectacular climax forest and mountain palisades.

The narrow collar of beach near the outlet was uniformly imprinted with fresh spoor from an extensive wolf pack. The largest set of prints revealed that two toes were missing from the right front paw. Greg wove an Indian broom of willow canes and brushed out every track, including his own, for more than a mile. Because there was no unobstructed passage from one end of the lake to the other, he quickly pegged and tied together eight dry logs for a raft he could pole along the north shore. Two days of downpour and two of bludgeoning gale cut perilously into his time budget. Persistent waterlogging increased the amount of daily effort required for poling. To achieve the lake's brushy west end, Greg paid a ridiculous price: ten days of devastating labour.

Time and labour worried him less than the presence of Trapper-Dan. After reaching a level, timbered cove with a broad, sandy beach, he set up base camp for an indefinite bivouac. The site itself, concealed in a stand of mature cottonwoods, afforded an unimpaired view of the lake and surrounding mountains. It would have been difficult – but not impossible – for an airplane to enter the upper five miles without being seen.

Fresh spoor, including the set with the missing toes, imprinted every damp surface of sand and soil.

On the morning following his arrival, Greg sat scanning the brushy ridge that formed an ampitheatre around the entire west end of the basin. A pack of timber wolves, trying to outflank the wind, raced single file over the ridge and into the northwestern watershed of a Skeena tributary. The leader had already disappeared by the time Greg first saw them, and so he had only one clue to associate the kennel with Náhani – two limping solid black individuals with thick mantles. They resembled the disabled pair that had always spooked when Greg walked upright near them at Nakinilerak Lake.

He rushed towards the hill. With all his strength he hurdled through knee-deep subshrubs, sphagnum, and devilishly spaced fallen timber to the stony ridge; but the wolves had disappeared by the time he reached a swale of open browrock from which he could study a wide sweep of terrain through the glasses. Tracking northwest, he jogged up and down ridge after ridge along what appeared to be an established runway until the spoor drove a straight line towards a buckhorn pass in the limestone range immediately south of Tatlatui Lake. That was the pass Greg had been unable to find during his flight from Tatlatui. There was no way for an Indian afoot to overtake a fleeing pack of wolves.

Hoping the troop was not Náhani's, he turned towards camp. Along the way he found a urine-marked shrub to which he tied a perspiration-soaked bandanna. He told himself that the wolves would surely double back along the same route, especially when Náhani got a whiff of firearms and airplane at Tatlatui Lake. She would recognize Greg's sweat signature and sniff his footprints into camp. To remove all the doubts from her mind, he took off his boots and walked the last mile barefooted.

That night, as he listened to the busy discourse from the woodland, Greg kept a small campfire burning until the morning star cleared the treetops. He played harmonica tunes the wolves had often heard at Nakinilerak. He crawled down the beach on all fours, threw stones into the bay, called, rattled cooking utensils, and whistled; but in the end finally concluded that such antics frightened away more wildlife than they attracted. Not a bear, fox, coyote, or wolverine dared clean up the twelve trout he gutted and placed on rocks near the wind-ruffled hem of the lake.

Before falling into a deep sleep, discouraged, he wondered: 'Have I gone and chased her square into Trapper-Dan's gunsights? How did Trapper-Dan know what old Moiso knew?'

*

For ten days following the sighting, Greg neither saw nor heard a wolf. August's Sagamore-Council-Moon was five nights into the waning. According to his logbook, this date was August 6. There were two full moons in the month of August, 1966 – August 1 and August 30.

No aircraft had passed overhead. His disappointment made him recognize – more than he had been willing to admit – the profound fascination he had sustained for that wolf. Hatred was an emotion previously foreign to Gregory Tah-Kloma, but subconsciously he was beginning to hate Náhani's persecutors.

The desire to have another look at Trapper-Dan's camp became an obsession with him. Good summer fishing and herbing had extended his dehydrated food supplies beyond earlier expectations; but even so, he had to grapple with the realities of weather and unexpected delays. On the eleventh day at the upper Thutade Lake camp, an inexplicable urge drew him back to Tatlatui Lake.

He packed his gear and left. First, he began the onerous trek along the ridge where he had sighted the wolves. His excitement grew when he reached the shrub where he had tied the bandanna. The handkerchief had been removed, but a thundershower had erased all tracks. By maintaining the same direction in which the wolves had disappeared, he walked straight to the buckhorn pass, a practical foot route through the precipitous range. At the end of two days, he reached the upper shoreline of Tatlatui Lake.

On the beach, it took no expert woodsmanship to determine that Trapper-Dan – and the wolves – had discovered his camp site. Boot prints and wolf tracks were everywhere, including those of the two-toed member. He estimated their age at ten to twelve days. The wolves had unearthed his campfire pit. His willow broom had been retrieved from the brush, where he had carelessly tossed it. Almost in panic, he counted seven widely scattered empty Magnum cartridges – all fired from the same rifle. Deep impressions from boot prints between the shells indicated that a running man had stopped seven times to shoot.

In Greg's reconstruction of what took place, Trapper-Dan and his men had either stalked or surprised the busy wolves in the act of unearthing the fire pit and exploring the camp site. The proximity of spent casings to the camp site, however, revealed that the ambush had been sighted by wolf sentries who alerted the pack before the man started shooting. The decaying carcass of a lame black female, skinned except for the feet, lay on the open forest floor 150 yards up the hillside behind the camp site – mute tribute to the lethal accuracy of the rifleman. Greg removed a foot from the dead wolf and put it into his rucksack. One of the Indians had driven a pointed sapling pole into the sward where Greg's sleeping bag had flattened the grass ... to put the young Chimmesyan 'in the shadow of the spear'.

When he reached the lower end of the lake Greg's suspicions were confirmed. The plane had taken off. The fire pit was buried. The size of a refuse heap indicated that the men had spent at least ten days waiting for the wolves. Greg was to ponder specific questions for the next several weeks: How did Trapper-Dan, Eugene Charley, or the white men know what old Moiso knew? How did they know the wolves would range the Tatlatui country at that particular time? Now certain that the pack was in the immediate vicinity, would the air-borne hunters be able to chart an established runway by plane? Was it as obvious as it had first appeared that Náhani was leading her troop north?

Perhaps the slaughter of the black member would force the wolves into life patterns exclusively nocturnal. At this longitude in August the nights were not six hours long.

Fatigued, tense, and threatened by heavy storm clouds, Greg stretched his rubberized nylon ground cloth over a double-A frame of sapling-spruce poles. Although he needed the rest enforced by the three-day storm that ensued, Greg begrudged the loss of time and the bite into his supplies. Prompted by an impulse the moment the wind changed direction, he scrutinized the aerial quadrant map, reached an enthusiastic decision, and started north even before the rain clouds broke apart.

Dense forest covered the terrain beyond the upland lakes, but the pavement of glacial till made walking easy. Because of more supple legs and shoulders, a light pack, and a strong hunch, Greg ambled along at an average of two miles an hour, deriving at last a sense of physical pleasure from strenuous use of hardened muscles.

At Kitchener Lake he glimpsed a small herd of woodland caribou. For fifteen miles along the shoreline he examined the tracks of a large wolf pack. Careful inspection of the distinctively broad lead prints – when they were not trampled out by the wolves that followed – confirmed earlier observations – two toes were missing from the right front paw. Either Náhani had met with mutilating misfortune or Greg was fruitlessly endeavouring to overtake a different wolf pack whose leader's paw print measured six inches in length. He had seen only one wolf print six inches long.

*

North of Kitchener Lake, in a region of glacial eskers and alpine subtundra, Greg drifted slowly for nine days in the wake of a northbound congestion of paw prints. On several occasions he lost the trail for as much as six hours. Did the pack sense a pursuer and purposely lead him over terrain in which even the sharpest-toed animals left no print? In any case, a large pack of twenty or more wolves left other signs for the Indian tracker: a turned pebble here, a clawed lichen there, scats, the smell of urine on 'boundary posts' along the runway, moult hair inadvertantly rubbed off against a tree, hardened summer resin 'clods' chewed from furry foot pads, the remains of prey.

In the Prudential Range there were nights when songs of distant wolves floated faintly through the air, but not from the bearing of any star Greg recognized. If he camped more than one night near dishpan lakes where he heard the serenades, either they sang no more or moved farther northward ... always farther north.

Softly one night he played the harmonica. He felt at peace. Even if he failed to overtake his enchanting Náhani – provided it was indeed Náhani's trail he followed – it was *he* who had actually succeeded in his mission. He was hounding her north into that region of the Kitiwanga beyond Trapper-Dan's and Eugene Charley's sphere of slaughter.

Then, high above the alpine lakelets whose overflow spawns the Stikine River, Greg received a disappointing answer to one of his questions. A pack of eighteen wolves of mixed age circled within fifty feet of his camp site one evening shortly before sunset. Their coats all bore the same family trait: bright brindled-grey. The adults were strikingly slender wolves, tall, sleek, efficient, and fast. Their coordinated trot reminded Greg of the piston movement in a powerful engine. Neither wild nor skittish, the intelligent, keenly curious troop sniffed, whined, then single-filed away without changing gait or speed.

Náhani had thrown him off her trail. Little doubt remained that she had led him into the established runway of the brindled wolves, none of which left a two-toed print. At some point along the way, she had funnelled her team's tracks into the runway, then turned her kennel over trap-rock and floatstone that paralleled the territory for a short distance, and then disappeared.

Greg began to consider the implications of this discovery. Supposing Náhani had no runway of her own! As renegades, supposing the wolves drifted from range to range, took what they wanted, and moved on in order to outwit pursuers. Such had been the life pattern followed by outlaws with prices on their heads. Those with greatest range and variety of habits survived longest. If her circuits of activity lay east of the Stikine headwaters, Greg might still have some assurance; if west, the bounty hunters would win sooner or later.

5 The Náhani mystique

Greg acquired fresh determination. Despite his hunch that Náhani ranged farther east and north, he left the brindled wolves' runway in the Prudential Range and grappled with an alpine route that led west above timber line through Skeena Mountain heathlands. By game trails he crossed the broad divide in the upper reaches of the Stikine Plateau whose ample water table provides the springs that become the Ross, Stikine, and Skeena rivers. For fifty miles he probed every clue along the runway of a large wolf pack until the trail was chopped into oblivion in a maze of wooded canyons that funnelled into the Nass River valley. Deer, elk, and bighorn used the trail so frequently that every canid spoor was effaced. Brushwood forest and rock-bound passes defeated Greg's every effort to reach the valley.

Although he carried a boot-repair kit, his footgear was worn beyond safety. No modern woodsman goes farther than his shoes. Therefore, when the Moon-of-Painted-Leaves brought the first colour to broadleafed foliage, he set a left-bank course down the familiar Skeena River canyon.

In order to avoid the settlements of Kuldo and Kispiox, as well as the extra mileage entailed in the Skeena's big bend, he undertook a little-used trail east of the Sicintine Range, west of the Shelagyote Glacier. Below the Babine River ford he hiked an old logging road and arrived in Hazelton shortly before midnight. He went directly to his friend, Rocky Longspear.

Before going to bed he revealed Moiso's information that Náhani had southed the Skeena the previous winter. He told what he had seen at Tatlatui Lake.

'I have to spend the winter around here,' he said. 'If

Náhani crosses the Skeena this winter, she'll get it sure as hell. I think I know the route she'll take if she comes down this way. Maybe I can rent a cabin from the Kuldo people.'

'She'll get shot if she goes to Kuldo again,' Rocky said. 'They'll be waiting for her this year. Like everybody else.'

'Then what about a trapper's cabin? Do you know of one I could rent? A wilderness cabin, north of here?'

'Only one.'

'Where? Who owns it?'

'There's a good one on Kitwanga Lake. Owned by Trapper-Dan Tall-Totem. He won't be using it. He'll be working all winter at the smelter in Kitimat.'

The next morning Greg took a decisive, if unimaginable, step. He went directly to Trapper-Dan's cabin where greetings were exchanged in the form of invectives. Dan's two squaws interceded before a fist fight could break out. They pulled both men into the cabin by their hair and stood between them. When the two agreed to sit before the log in the fireplace and smoke the calumet, the squaws said they would prepare a big meal of grilled salmon, hoecakes, and strong coffee.

After the meal Greg spread his maps on the oilcloth-covered table and traced his summer route.

'I ain't seen nobody but a Tsimshian trapper cover that kind of ground in the Kitiwanga,' Dan said. One of the squaws combed his long, coarse hair. 'If I hadn't seen that Vibram design in your boot prints at Tatlatui, I'd deny you done it.'

'Why did you kill the crippled black wolf instead of Náhani?'

'You must've come back to know that, huh? You left when you saw our camp and the plane. That yellow Eugene Charley could've killed Náhani. Had her in his sights and then his liver turned to cloud-milk. She'd already gone over the hill when I got there. Now what?'

'I'm going back next summer,' Greg said. 'I'll spend the winter near here if I can find a cabin for rent. The Chimmesyans say she'll come south for the winter.'

'Did you communicate with that wolf?'

'Maybe.'

Trapper-Dan's attitude became almost servile. He explained that he owned a fine cabin near Kitwanga Lake, thirty-five miles northwest of Hazelton. He confirmed that he had a winter job at the smelter in Kitimat. No use trying to trap until he could kill that wolf. He offered to rent Greg the cabin and help haul in winter supplies. He spoke of another empty cabin at Swan Lakes. He was willing to take an oath that Tsimshians had sighted Náhani near both lakes.

Trapper-Dan drove his Land Rover and Greg followed in his station wagon to Smithers, where they loaded the two vehicles with an eight-month supply of food, winter clothing, candles, and other staples. The merchant sent an order to Prince George for a four-month supply of dehydrated items and a new pair of boots to be picked up sometime in May.

It took four trips – two days and one night – for the two men and the squaws to pack 800 pounds of supplies and equipment six miles beyond road's end to the cabin. Trapper-Dan refused Greg's offer to rent mules. Said he owed the young Chimmesyan an apology for the way he had tried to kill Náhani. If his job at Kitimat panned out, he'd give up trapping altogether.

By candlelight Greg wrote in his diary: 'When I let it out that Náhani might come south this winter, and that I wanted to rent a cabin somewhere north of Hazelton, all the hostility and meanness vanished from that bastard's face – a face so damned full of ugliness you can hardly tell when it changes expression. Right away he offered me all this help – even a young squaw for winter shack-up if I would feed her. Butter wouldn't melt in his mouth. Why? I wonder what cards he is holding and how I'm playing right into his hand. Why did he want me to hike sixty miles to Swan Lake and back?'

The 'fine' cabin proved to be primitive and in bad repair; but its eyrie location, with western exposure high above an east beach, and picturesque little food cache on four animal-proof poles, made up for inherent defects. Greg looked forward to uninterrupted weeks of solitude during which time he planned to cultivate the friendship of native wildlife, and build the physical fitness necessary to penetrate the northern Kitiwanga.

In chilly October he repaired the one-room cabin's structural dilapidation, rechinked the wall logs with sphagnum and sapling battens, caulked the flooring, and weather-stripped the double door. Both windows had excellent storm shutters, and the split-cedar shake roof required little attention. He sawed, split, and carried twelve cords of quartered spruce, fir, and cedar from nearby deadfalls.

There was no stove. Cooking took place on a fireplace grill. Water came by bucket from a small creek dam fifty feet from the door.

On the tenth of October, Trapper-Dan and three husky Tsimshian bandsmen appeared. They laughed when Dan offered to bet that a Chimmesyan might survive the Storm-Chief-Moon and one or two Hunger-Moons. Trapper-Dan insisted upon leaving a ·30–30 and a box of cartridges.

'Lotta grizzly in here. They break in even if you store your food in the cache house.'

One of the bandsmen left a fine pair of hickory and moose rawhide snowshoes. The men ran back down the trail within five minutes of their arrival.

Greg sensed in their behaviour an ulterior motive that he

failed to decipher at the time. Trapper-Dan was a reservoir of unpredictability. The gun and snowshoes had to be some kind of plant.

As Autumn advanced, more unexpected visitors found their way up the trail to the cabin. One afternoon while Greg was splitting wood, a middle-aged Tsimshian – a slender wedge of a man with a grooved and wrinkled face – introduced himself as E-Leish, son of Moiso.

'You left for the Kitiwanga while we were in Smithers for supplies,' he said. His furrowed lips trembled as he spoke to the 'wolf-man'. Because individual pronunciation varied so much, Greg was often uncertain whether the bandsmen referred to the village of Kitwanga, Kitwanga Lake, or that mysterious plateau country they generally pronounced *Kee-tee-wahn-gah.*

With intense interest and unflinching gaze, E-Leish studied Greg's eyes as the young Chimmesyan explained why he had been compelled to struggle through more than 300 miles of pathless wilderness during the past summer. Greg was aware of a childlike frankness and naïveté in the man who obviously considered everything he didn't understand as preternatural or shamanistic. He wore a necklace of adamants, moose teeth, salmon vertebrae, and wolf claws.

'My people are Wolf clan,' he said. 'We carve totems. Trapper-Dan he don't believe in totem. Too much booze. Tsimshian people don't trust Trapper-Dan.' He looked up and down the trail before continuing. 'Last winter bad. In the valley of winding water my house is next to my mother Moiso. I have one window. My squaw and me sit at the table one night. We listen to the Storm-Chief. Celia, my squaw, grab my arm. She look at the window. There stand Náhani. Náhani blow breath on the window. It freeze like rime. Melt only in May. Her face shine like the night sun.'

He went on to say that a missionary and a sachem had explained the phenomenon by declaring him under the spell of a devil.

'The white man's priest knows more about heaven than he does about Indians,' Greg said. 'Your sachem knows more about Indians than he does about wolves.'

For an hour Greg sat on the log and related his experiences with Náhani.

'There is nothing supernatural about that wolf,' he concluded. 'Normal in every way. I've walked many miles with her. Sat with her night after night at the campfire. Eaten with her. Slept while she and her pack sat near my sleeping bag. I knew her only as a wild animal – but as a friend. I saw her care for her wounded and defend her kind against Ozilenka, the grizzly. Of course she kills her own prey, as all predators do. Nothing more. Wolves don't attack people. If she looked through your window, it was natural wolf curiosity – maybe good medicine. I am here because I want to associate again with Náhani.'

E-Leish took from his haversack six long guts filled with pemmican, which he placed on the wood stack. Greg went to the cabin and returned with the dried foot he had cut from the dead black wolf at Tatlatui Lake. He gave it to E-Leish. When they shook hands a wistful look came over the man's face – but the look of oppression was still there.

Two days later Greg was reinforcing the cabin's outer door. A man about thirty years old and a squaw appeared on the trail.

'They call me Peter after my uncle, Buffalo-Bones,' the man said. 'You knew his son, my cousin, No-Guns. This is my wife, Lorna.'

'Peter,' Greg said, 'I thought I was seeing a ghost. You look so much like No-Guns.'

No-Guns, a boyhood friend of Greg's, had lost his life in a

lumber-mill accident. Peter and Lorna were a handsome couple, open-faced, easy-smiling, naked-hearted, like No-Guns. Peter was tall and slender like his cousin. After coffee and the usual amenities, he and Lorna came to the point of their visit.

'News travels fast up here, Greg,' Peter said, as if what he was about to reveal pained him. 'Tom-tom telegraph. Council smoke around Babine Lake says you'll winter here in Trapper-Dan's cabin. A lot of Indians still depend on trapping in these woods, Mr Wolf-man.' They all laughed, but the mirth was strained. 'I suppose you know what Náhani did to the trap lines the last couple of winters. Bear Lake people had to leave Driftwood Valley because the pack followed every hunting party for two Hunger-Moons and drove off the game. Arch Aloyet got a close shot at Náhani near Takla Landing. He's the best shot in the province. Said his bullet went clean through her as she charged. Before he could shoot again, she threw her weight against him, grabbed his rifle, and ran off with it. The same wolves were seen taking livestock as far south as Burns Lake. I hate to tell you this, Greg, but the reward now stands at a thousand dollars. It'll go to fifteen hundred if she souths this winter.'

'How much of this malarkey do you believe, Peter?'

'Hard to say. The main thing I came to tell you is that a no-good Carrier son of a bitch named Eugene Charley rumours it around that you plan to bring Náhani and her pack to Kitwanga Lake. That's why they beefed up the reward. Talk says the RCMP may take you in. Trapper-Dan thinks Náhani will find you here. He says you were with her last summer. Says you know how to talk with wolves. He wants more than that reward. He'll shadow this cabin every moon. And so will Eugene Charley. That's why Dan rented it to you. If Charley gets Náhani first, Dan'll do him in and claim the reward. Don't trust either man, Greg.

One of them will break your bowstring ... once Náhani's dead.'

Three days after Peter's visit an RCMP vehicle towing a horse trailer stopped at road's end. Constables Larson and McIntyre rode up the trail to the cabin.

'Are you the guy that hobnobs with wolves?' Larson asked. He had the eye of a comedian and expected you to laugh either at him or with him. 'Do you realize what stink you've stirred up with this Náhani? Are you really in cahoots with the outlaws? You look rather sane – for a Chimmesyan, that is. We expected to find a real kooky bloke – maybe a hippie. Do you want to tell us about yourself – and your playmates?'

Over coffee and a popping log fire, Greg related every detail of the Náhani encounter as well as the reasons why he planned to spend the winter in the cabin instead of Prince George – why he planned to carry on the search the following summer.

'Good Lord, Tah-Kloma!' McIntyre exclaimed. 'You're even kookier than we thought. We'll have to keep an eye on this situation. If she shows up, we have only one choice. Too many people stirred up on both sides of the fence. Feuds start that way, don't you know?'

*

A week went by. Wolves on the ridge behind the cabin had sung for two nights. And then on Friday afternoon Greg recognized the bent and staggering figure of Eugene Charley trudging up the trail, alpenstock in one hand, carbine in the other. Greg hoped he could resist the temptation to pound Charley's nose. The Carrier was fully equipped with tumpline basket pack for fall hunting. His new jeans, shirt, and jacket, purchased the previous season, had not yet been to a laundry.

'I rode with a logger from Babine Lake,' Charley explained. 'Trapper-Dan brought me in the Land Rover to the trail.'

'Trapper-Dan is working at the smelter in Kitimat,' Greg said with a smile that accused Charley of a forked tongue.

'We go to work next week. Dan said he'd be here Tuesday with more supplies you'll need. Stuff he forgot last time.'

Suddenly Greg had an idea.

'I'm leaving tomorrow morning for the Swan Lakes. You can keep me company. I think we'll meet some wolves. There are two cabins, fifteen miles apart.'

'Thirty miles!'

'Sixty, round trip.'

'So you made friends with that damned wolf at Nakinilerak. And last summer at Tatlatui?'

'You and Dan know she is coming south, don't you, Charley? You know I spoke her tongue last summer, huh? Dan told me how you buck-eggered instead of shooting her at Tatlatui. I found your spear. You were born under a falling star, Charley. I want to see how you act when you come face to face with Náhani at Swan Lake.'

A catlike grin crept around Charley's weathered lips. 'You'll see.'

Greg was surprised to find Eugene Charley ready the next morning for an early start to the Swan Lakes basin. With medium-weight packs, brisk November weather, and a well-beaten trail, they travelled fast once beyond the ridge above Kitwanga Lake. The route alongside the Cranberry River traversed luxuriant conifer forest. Stripped of their foliage, riparian hardwoods now stood gaunt and inflexible. The last vestige of early autumn haze had blown away on polar northers. Distant Mounts Thomlinson and Tommy Jack, already frosted crystalline white, projected themselves against the polarized segment of the sky.

How that Carrier could walk! Greg had to run to keep up with him. Charley was accustomed to long treks with a pack load tumped across his forehead. No fifteen-mile jaunt could faze him.

Without difficulty they found an empty cabin near the junction of four well-used trails at the base of Mount Weber's north shoulder. Even after his slug of rum that evening, Charley was far from his talkative self. He insinuated that Trapper-Dan was cooking up a plot to do away with him as a potential beneficiary of the huge bounty on Náhani. What he really feared was that too many Tsimshians were aware of his duplicity. He was ill at ease in Tsimshian country. Moreover, he had delved sufficiently into Náhani mystique to suspect Greg of leading him into a trap. Even as he slept he never relinquished a knuckle-whitening grip on the carbine.

The next day as they crunched along the frosty trail up the Kispiox River valley Greg felt momentarily sorry for Charley, an Indian so little at peace with himself and the Earth Mother. He had hated the Carrier after seeing him at Tatlatui Lake and knowing the extensive lies he had circulated in order to get the bounty raised; but now he recognized him for what he was, a captive of negative appetites, none of which he could control – rum, greed, and a passion to be someone he was not and never could be.

After crossing a low hogback, the trail swung down a pine-wooded slope to span an open savannah and arrive at the sprawling, irregular eastern shoreline of the first Swan Lake. On a slight rise fifty yards from the beach, someone had built a solid cabin with heavy, unpeeled cedar logs, split shakes, and adzed fir flooring. There were two windows with shutters properly hinged and hasped.

The two men ate supper without speaking. The ominous silence was broken only by the hiss of a rising north wind.

Shortly before midnight, under low-hanging nimbus

that screened the waning Fallen-Leaf-Moon, a chorus of wolves sang near the lake shore less than 100 yards from the cabin. Eugene Charley swigged a whopping draft of rum and cursed Greg for leading him into a trap. He ran to the door and held a flashlight alongside the front sight of his carbine.

'Náhani!' Greg shouted at the top of his lungs.

Charley fired seven times in the direction of the lake. Then he did exactly what Greg hoped he would do, what Trapper-Dan hoped he would do. After he shot at the wolves, he stuffed his belongings in the tump pack and left.

*

Some time that morning Greg fell into a deep sleep. He bolted upright when the entire cabin shook with an explosive boom. The blastlike wind had broken a heavy dead limb from an overhanging conifer, and the 'widow maker' had crashed through the cabin's roof ridge. Dark clouds swirled above the treetops. When the droning wind began to drive fish-scale snow horizontally across the basin, Greg hit the trail for Kitwanga Lake.

He had climbed to the first hillside, still within snow-veiled sight of the cabin, when he heard the hunting bay of a wolf pack. He turned and listened. Eighteen to twenty-four wolves, using the horseshoe hunting formation, converged upon the cabin and circled slowly as they do when they 'set' a prey animal at bay. Suddenly the leader wolf, a large silver-white animal, discovered Greg's foot scent trail. The pack broke into a dead heat across the open savannah.

'Náhani!' Greg shouted as he ran to meet them. They were less than a hundred yards away and racing towards him. Sound, like vision, becomes distorted in a swirling snowsquall. He saw the wolves thrust forward their long legs in order to stop and sample the air. He called over and over again. For a few moments it looked as if they would allow him to approach. Then they turned in a body and scampered away behind their leader along the frost-hardened shoreline of Swan Lake.

Greg could not risk becoming snowbound in that cold cabin without supplies or the means for repairing the roof. Reluctantly he resumed his journey, not certain he had seen Náhani, but apprehensive for fear he had.

6 A fragile image

Wind at first kept the snow from drifting and covering the trail, but the storm intensified. When Greg dragged his feet up to the Cranberry River cabin around midnight, a faint glow of orange-red candlelight shone through an unshuttered window. Sparks from the chimney revolved in billows of smoke that puffed into the snow-laden wind. In the eerie snowlight the cabin's frosted north side looked as if it were carved from a frozen cloud. One glance through the iced south window disclosed the blurred outlines of four people. The frosty glass made recognition impossible. Three were sitting at the table on which flickered a single candle. The fourth was tending the fire.

'Hello!' Greg shouted and rapped on the window pane. The four made no move to answer. He walked to the door, knocked with his alpenstock, and shouted again. Finally the door creaked open about the width of a finger and someone muttered in broken English: 'Go away, Chimmesyan wolf-man!'

'I can't get over that pass in this storm. Let me in.'

'No. Go away, wolf-man.'

The door slammed and was bolted with a drop bar from the inside.

'I'll wring somebody's ruddy neck for this!' Greg shouted.

With new flashlight batteries, he had no trouble staying on the path near the Cranberry's right bank; but at daybreak when he reached the north foot of the Kitwanga ridge, he found that the mile-long drifts had buried the trail. He mushed out the day, backtracking from one cul-de-sac to the next. He was too busy and too tired to worry about the four men in the cabin and their intentions. They knew who he was. Their lack of hospitality had been deliberate.

By the time he broke a route around the east prow of the ridge, he realized that his uncharted wandering had taken him five miles below Kitwanga Lake. Darkness and heavier snowfall blanked out the landscape. The brush grew too thick for the angle back towards the lake, so he traipsed out over a frozen marsh to road's end.

There stood Trapper-Dan's Land Rover. It was locked tight or he would have spent the night inside the vehicle. Could the four in the Cranberry River cabin include Tall-Totem? Did those men expect him to lead them that easily to the queen wolf? Someone could have 'borrowed' the Land Rover while Trapper-Dan wallowed in a drunken stupor.

Finally he reached the cabin. No one was there. No one had been there. With an extravagant meal in his stomach, Greg fell asleep in a chair in front of the fireplace. He awakened when a whisky-jack tapped a familiar code on the windowpane. The reed-voiced bird received an immediate reward of peanuts for the warning he gave.

The sky had cleared. Frosty blue. No wind. The jabber of Tsimshian dialect and the leathery crunch of snowshoes reached Greg's ears long before he sighted four men hurrying from the direction of the Cranberry River trail. Inasmuch as the snow had covered Greg's tracks during the storm, the men had no way of knowing he had not perished in the blizzard. Prepared to clobber noses one by one when they entered the cabin to split up his supplies and gear, he was astonished to see them crunch on down towards road's end without so much as a side glance. They were too heavily bundled in parkas and blankets, too bent over, too darkly goggled for recognition. Not one of them, however, was tall enough to be Trapper-Dan. Each man carried a gun.

Forty-eight hours later two RCMP patrolmen with packs

and scoped rifles arrived on snowshoes. Again it was the whisky-jack's windowpane tapping that brought Greg to the door.

'We heard you and Trapper-Dan lugged all your supplies up that trail on your backs,' said the sergeant named Duncan. 'I can't believe Dan Tall-Totem didn't rent a string of mules to do the job, since you were footing the bill.'

'Dan claims the smell of a mule's hind end retches him,' Greg said.

After they had finished dinner and arranged the two sleeping bags near the fireplace, Greg outlined his trek to the Swan Lakes and return. He omitted only that part where he thought he saw Náhani.

Duncan spoke. 'I think you recognize, Tah-Kloma, that we are here because of a report that a large wolf pack did pay you a visit at the Swan Lake cabin. Was the silver-white Náhani in the pack?'

'If she had been, I believe she would have come directly to me. I called, but they disappeared like thistledown in a snowstorm.'

'We know,' said the corporal named Blakeburn, 'that the four Indians are working for Trapper-Dan. They knew all along about the pack. They suspected the presence of Náhani. They followed Charley and you and were hidden in the forest, hoping for a close-up shot. Under the influence of rum, Charley blew their chances when he rushed out and fired his thirty-thirty. The weather also loused up their luck. Trapper-Dan, of course, wants us to believe *you* killed Eugene Charley.'

The two mounties left the next morning for the Swan Lakes. They planned to follow the Nass River trail back to Terrace unless they unearthed 'something interesting'.

*

During November's Moon-of-the-Storm-Chief, Greg took advantage of 'a mild spell' in order to catch and smoke a hundred pounds of trout, kokanee, and char, which he stored in the weatherproof food cache before December's Moon-of-the-White-Hare freeze. There was always more to do than there were hours in which to do it. Kindling to split and dip into a bucket of melted pitch for log lighters; an increasing number of non-migrators at the food trays; household chores: laundry, cleaning, repairing; hikes to distant ridges to maintain a level of physical fitness; careful diary and logbook entries as free as possible from invectives and libel.

Autumn was late and mild. Misty rain melted earlier snow, leaving the clearing around the hillside cabin an island of mud. One morning Greg found wolf tracks. A large pack had soft-padded within two feet of the steps, but not a breath had alerted him to their presence. The day was clear, so he hounded the prints down the trail. Half a mile before road's end, they led to a bloody blotch and the last of a set of elk tracks. The wolf spoor disappeared towards the hills above the riverine valley of the Kispiox. He slogged along the muddy trail to road's end to check out four distinct sets of boot prints that led to and from the site where the elk had fallen. Men had visited the scene before the wolves. Wolf prints covered boot prints.

He had no sooner returned to the cabin when an RCMP float plane landed on the lake and taxied to the beach. Corporal Blakeburn and the pilot emerged.

'Anything new?' Blakeburn asked.

'I just got back from road's end,' Greg said.

'Rumour has it that your Náhani is along. The pack killed an elk down about road's end. Log fellers made positive identification. I thought you said she wintered in the Kitiwanga and chased caribou.'

'Could be she changed her mind.'

When they reached the cabin there was no use trying to conceal the fact that a wolf pack had been there. Blakeburn's suspicions surfaced as if he had just impeached Greg's integrity.

'Rendezvous?' The tone of his voice was filled with icy insinuation.

'I was asleep. Never knew when they were here. That's why I followed the track to the elk kill.'

'Which way did they head?'

'Into the Kispiox. The only other tracks were made by the elk and four game poachers. I checked the tyre tread in the mud at road's end. Land Rover. The elk was dead and gone before the wolves got there.'

'We thought you ought to know about the big weather front moving down from the Gulf.'

'That's why the wolves came to the cabin. She tried to warn me. Tomorrow they'll be in the canyon of the Skeena.'

'A hundred fifty miles?'

'No trick at all under her leadership.'

Blakeburn handed Greg a pound of pipe tobacco, two dozen candles, a can of plum pudding, and six cans of condensed milk from a small haversack he carried.

'That's for one-man potlatch,' he said. He smiled for the first time. 'Sorry we have to run. You can pole us off the beach. Carry on!'

For the next five days and nights Greg left the cabin only when necessary. Twenty-four hours of sleet and rainfall left crystalline stalactites a foot in diameter between the cabin eaves and the ground. As the wall of arctic air moved in, the rain turned to snow. Greg shovelled for two hours daily to prevent heavy accumulation around the cabin. Between flurries and wind blasts, a freezing fog – 'frost smoke' – arose from the marsh during the long transaction of freezing.

Then the lake began to freeze. Growing ice crystals

groaned and creaked as they extended from shoreline. They shoved and crowded one another for space at the surface. When the temperature dropped to twenty degrees below zero, groaning and creaking gave way to explosive blasts as expanding subsurface crystals demanded space.

The storm's smells brought new texture and chemistry to the air. Miles of forest on both sides of the lake swayed like wheat fields under the wind's onslaught. Soon a plating of snow frosted the surface of the woods, and for two Hunger-Moons the world remained solid white. Balsam and ozone permeated the air.

Occasionally Greg could see across the lake. At other times visibility was restricted to a few feet. Once he saw his breath form a cloud ten feet long that drifted all the way from the cabin to the lake. Out of doors he plugged his nose to keep from breathing through it. Day or night he could estimate the extent of temperature drop by the way trees exploded like shotgun blasts.

During that first storm the wind caused the fireplace to gulp extravagant quantities of firewood. Little heat remained in the cabin. Greg spent more time in the sleeping bag and caught up on all the sleep he had lost during the preceding two years.

Until the temperature moderated enough for wind to crust the cotton snow, his movements were limited, even on bearpaw snowshoes. The first task was to tamp a trail to the creek where ground grue slowly displaced the last running water. For a month he kept two buckets of icicles melting near the fireplace, but when the supply was exhausted, he resorted to snow for drinking, cooking, laundry, and the quick once-over he considered a bath. For a while he melted frost rime from ice-plumed high-bush cranberry canes, but the second storm buried the fenlands beneath twenty-foot drifts.

*

All through the December and January oestrual period, mating duets of wolf pairs drifted down from the highlands, especially after January's Big-Sleep-Moon waxed full on the twenty-sixth. After breeding, the pairs regrouped into packs to resume organized hunting. Kitwanga Lake wolves operated in small family groups. They carefully avoided the cabin vicinity with painful recall of Number 4 steel traps, gunshot wounds, and the stench of death during Trapper-Dan's occupancy.

No creature at Kitwanga Lake demonstrated more persistence than the short-tailed weasel or ermine. While January winds knifed across every open expanse, Greg saw the small creatures waiting for hours against the snow near openings in muskrat lodges. Since muskrats left their lodges to excrete, the ermine knew it was just a matter of time before his uncomfortable victim would be forced to emerge. There was generally a fight over the carcass, because short-eared owls, concealed on branches above the frozen shorelines, awaited each kill. Occasionally two weasels clung to the lodge, in which case one ate muskrat, one ate owl. Sometimes owls took both weasels, then a hidden mink rushed out for the muskrat; but if a marten joined the drama, he might dine on mink, muskrat, or owl. The food chain was composed of many links.

One moonlight night when the mercury in the white enamel thermometer outside the door hovered at forty degrees below zero, Greg heard the leathery crunch of approaching footsteps. Nearby wolves had just finished a noisy ruckus in the forest behind the cabin. Expecting to greet a human visitor, Greg rushed through the storm door; but when he stepped outside under the broad A-frame overhang the 110-degree temperature difference caused his face and eyes to feel such pain that he suffered a temporary blackout. Ordinarily, when he went out at night he was careful to accustom his exposed face and lungs gradually to the debilitating temperature change. Unless he wore a fog mask, his own breath cloud after dark obscured visibility almost as effectively as the whiteout of winter brume. On this particular night, he stumbled and groped back inside the cabin's inner door. For a time he feared that the corneas of his eyes were frozen.

As he was recovering, he heard a deep-throated woof and gargle. It was then he realized that the wolves had discovered and routed a hibernating bear very near the cabin.

Quickly he put on his down-packed nylon parka hood, attached the fog mask across his mouth and nose, and with gauntlets strapped below the elbows and mukluks buckled, went outside. A black bear reared and moaned from the cabin's top step and tried to shoulder herself inside; but Greg led the way to a crawl hole under the south side of the cabin floor. Trapper-Dan had built the dwelling on a hillside. The four-foot space under the front between ground and floor joists served as tool shed and catchall. The clever north wind always managed to sneak through the flooring, so Greg had stuffed the area solid with a winterizing insulation of sphagnum. When the bear saw him digging out the dry moss, she plunged through the crawl hole, snarled, and buried herself in the spongy material. Greg plugged the hole with snow and moss so that the bear might sleep undisturbed until April's Moon-of-Leaves-and-Flowers.

Unable now to harass the bear, the wolves pursued their never-ending games of tag on the western shore of Kitwanga Lake. On clear evenings when the wind modulated, Greg followed the pack. Tamping a path with broad snowshoes over the snow-covered surface was good exercise. At minus fifty degrees he had to keep moving, regardless of how well dressed and warm he felt. But each time he began to perspire, he slowed down; otherwise, his underclothing would have frozen to his body.

On darkest nights he could approach the wolf pack to within fifty or sixty feet before they bolted away. From lake centre he looked up to see the aurora borealis crackling and dancing above. Chimmesyan sachems referred to the northern lights as heavenly gardens where the roses of yesterday bloomed again. But for Greg the combination of wolves and the aurora was a painful reminder of the many nights he had watched the spectacle with Náhani at Nakinilerak Lake.

Towards the waning of February's Moon-of-the-Fallen-Antlers, Greg began to feel a physiological and psychological rebellion against the interminable hostility of winter. Restlessness overcame him as he thought of the long months he still must wait before he could resume his search for Náhani. Despite his love for winter's moods and his respect for the season's realities, the restlessness remained.

Under March's Moon-of-Meltwater the wind often howled in concert with meat-craving wolves; for March, like the preceding three months, was also a Hunger-Moon – by then all the *easy* prey had been taken. Falling snow still kept the land crouched in icy paralysis. When cloud-shafted sunshine finally broke through, Greg ran out and threw snowballs towards the lake. As the last snow sifted down, he sensed a change in the final silence. He rushed to the thermometer.

Ten above!

The Old-Man-of-the-North was on the run! A warm Pacific chinook chewed at his heels. In the first sunset of spring lived the Bear that blew the west wind.

Greg spent most of March and April with the maps, and laid out a day-by-day itinerary from June until September. He worked for hours on every item of equipment, modifying each unit according to what he had learned the previous summer. To regain physical stamina lost indoors while the storms raged, he scheduled long hikes uphill and down, with four and a half stones in the Trapper Nelson. Spring rain had slushed the corn snow on southern exposures, but elsewhere the pack was still firm. He found it necessary to wear snowshoes after the sun had shone on the drifts for a full morning.

Dark cracks in the frozen lake opened with a saw-toothed roar. April's Moon-of-the-Flowers waned before spring finally seized the land. The old female bear dug out of the sphagnum and headed for meltwater. She gulped deeply from the creek that had begun to trickle. After urinating, she returned to the cabin. Greg handed her a smoked trout. Still plump and in prime winter pelage, she took the fish and headed for the forest trail behind the cabin.

Greg followed her for about fifty yards inhaling the softer scent of spring.

It was there, along the muddy trail, that he saw the fresh boot prints – large prints – one Vibram, one British sole rubber. They led to a clump of budding aspens about seventy yards from the cabin. The men had walked back and forth as if trying to reach a decision. The tracks then led up the hill, cut back in a wide arc, and finally re-entered the trail for the return to road's end.

Greg had been over the same ground the day before, and had seen no human prints.

7 The danger route

During the firelight hours in early May, Greg became aware of human shadows in the brush near the lake and in the forest behind the cabin. When he investigated by day, he found more boot prints in slushy snow and mud. There were four pairs of tracks, making the round trip to road's end. One set of familiar tyre treads printed the muddy road.

On the evening of May 10, he decided to bring the situation to a head. He fired Trapper-Dan's ·30–30, aiming shots at tree trunks close to the shadowy movements, and exercising care to hit no one. He fully expected return fire, but received none. The shadow men simply vanished.

He hated firearms, with their aura of violence and tragedy, but realized the need to keep the weapon close at hand. When his daytime activities were completed, he loved to sit on the top stoop-tread, lean back against the cabin door, smoke his pipe, and watch sunset colours glow on the lake, sky and western crags. Twilight – owl-light they call it up on the Ishkheenickh River – was his most treasured time, and he resented the sinister presence that disturbed his mood. He counted the days until he could know freedom again in the vast Kitiwanga.

Meanwhile, the snow line receded. Everywhere the earth awakened and moved. Buds swelled, and animal young ventured forth. Restless migration began even before the game trails were free of snow and waterways clear of ice. Clouds, fog banks, and vagrant chinook winds – even the chill wraiths of dawn – seemed to compound spring's restless motion.

As the earth moved into the vernal equinox, the air filled with swirling wings and strident song. Each evening, wedges of geese settled on to the lake for the night's rest before continuing the long flight to the Arctic tundra. Chimmesyans say the fairest of all springtime motion in the north woods is waterfowl home-coming.

Wolf packs also joined in the frenetic movements of late spring. Tsimshians believed that wolves from as far north as the Yukon had southed because of the severe winter. An unusual number of prints had appeared even before January's Big-Sleep-Moon.

There remained no doubt as to Tall-Totem's reasons for renting the cabin to Greg. As an Indian, Dan may have reasoned that Náhani would lead her pack south in order to take easier prey than herd-guarded caribou in the Kitiwanga. Another incentive for wintering nearby was the wolf requirement for certain greens after whelping: bluejoint, vernal, sorrel, wheat grass, and brome, which, because of late snow-melt farther north, would be available only on southern meadows in the Skeena and Babine river valleys.

Each time fresh tracks showed up at the cabin, Greg discharged the carbine, knowing that if Náhani were nearby and associated the weapon with him, he might never see her again; but if gunfire would frighten her away from a situation that endangered her life, he had no other choice.

On May 15 a group of Haida Indians from the coast drove to road's end, then hiked to the cabin. Their clan effigy was a silver wolf. They had already heard the saga of Gregory Tah-Kloma and Náhani. Rumour, they reported, was now rampant that rabies was responsible for the presence of so many wolves south of the plateau country. They also added that the RCMP feared Greg might be a carrier of the virus because Náhani had chewed his legs and arms. Veterinarians predicted a rabies epidemic among canids, wild and domestic. They also reported that Trapper-Dan had been interviewed by a newspaper reporter; he had declared categorically that he had seen Náhani's pack run with typical rabies falter, that he had shot one of her wolves, one that staggered so seriously he was unable to keep up with the pack. The coast Indians prayed that no harm would befall Náhani.

For a long evening after the Haidas had gone, Greg pondered the question of sylvatic rabies. He recalled an episode from his childhood. A strange-acting lone wolf, shunned by his kind, had wandered to the family cabin on the Ishkheenickh riverbank one spring. Although the wolf had no fear of humans, he was weak, emaciated, and irritable. Greg's father, Youngpine, had called attention to the animal's terror of water – an ominous symptom – and had noted that the wolf was ostracized by other wolves that were occasional visitors at the Tah-Kloma home. Within three weeks the wolf had attacked horses, deer, and coyotes. One morning Greg's father had found the animal near the doorstep – dead. Youngpine then shot every mammal in the vicinity. He explained that a carrier of sylvatic rabies might lead a normal life for years before succumbing to the morbid state.

In his own mind Greg was reasonably certain that no member of Náhani's pack was a rabies carrier. They were all eager swimmers in the lake – hydrophobic wolves avoided water. But the mere hint of a rabies epidemic, coupled with Náhani's reputation, generated unreasoning hysteria and panic among Indians and whites alike. The reward was soon to soar.

*

On May 23, 1967, Greg packed his essentials into the Trapper Nelson and headed for Hazelton. Twelve hours later, under the full Moon-of-Fawns, he arrived at the home of his friend Rocky Longspear. After outlining his plans for the summer search, he drove to Smithers. On the twenty-fourth he picked up a four-month supply of freeze-dried and dehydrated staples. He returned to Hazelton at midnight.

'I think one of Trapper-Dan's men was in the trading post and recognized me,' Greg said to Rocky. 'The guy got in one hell of a hurry to leave.'

'This fellow, Trapper-Dan Tall-Totem, is a bad egg,' Rocky said. 'Never been any good. Tsimshians are ashamed he is one of us. He has four, five friends he bullies around – drunkards, gamblers, thieves, all of them. The rest of us don't know what's going on, because we don't speak with men whose tongues are forked. Dan was soon fired from the smelter in Kitimat. Mounties have driven to his cabin many times lately.'

'What's the latest?'

'Everybody up and down the Skeena is on tenterhooks because Tall-Totem linked Náhani with an outbreak of rabies. You'll find Tsimshians unfriendly, Greg. Most everybody believes you killed Eugene Charley. That's why Tall-Totem left you a gun at the cabin. He figured you hated Charley enough to kill him. Tsimshians are not happy because the Haidas are sticking their flat noses into our business. The reward for Náhani now stands at eighteen hundred dollars. Come the Moon-of-Painted-Leaves, don't you come back here. I don't know that country, so I can't advise a trail out. Get to the Alaska Highway somehow. And call me on the phone. I'll bring your car. But don't come back here.'

Greg knew his friend well enough to listen attentively.

'One thing's for sure,' Rocky continued. 'Tall-Totem and his trappers will follow you this summer. You will be lucky to escape them. So don't go up the Skeena Valley like you did last summer. Go up the Kispiox. I'll drive you as far as the road goes – about forty miles north of here. Don't build fires for two, three days. I don't know how you'll keep from leaving tracks those human bloodhounds can't follow. They can out-track a hound. Stay off the trails. Keep on the ridges. Keep your eyes peeled.'

'That Kispiox River Road . . . how can we pass Trapper-Dan's cabin without somebody seeing us? By now he knows I've left Kitwanga Lake.'

'It's a chance we have to take for a forty-mile head start. It's two a.m. Let's go.'

'In about a week, Rocky, send word up to Dan that I left everything in the cabin in better shape than I found it. I want no more dealings with that guy.'

Rocky drove Greg to the road's end about five miles south of the Swan Lakes trail. After wading the shallow Sweetin River ford, Greg hiked near the path – but not on it – along the left bank of the Kispiox River. The route led directly to the Swan Lakes cabin. He decided to rest a full day and night in that vicinity, perhaps the last likely area to which the trappers would expect him to go. It was easy enough to walk on supple duff within twenty feet of the trail and leave no obvious boot prints. To further confuse the Tsimshian trackers, he wore heavy moosehide moccasins over his boots.

At dawn the sky looked as if rain would fall within the hour. He erected a hidden shelter on a brushy knoll a mile above the Cranberry-Kispiox trail junction. The site was a natural observation post, affording views of both trails.

Because it was still raining twenty-four hours later, Greg shouldered his heavy gear and struck northwest through gravel eskers until he reached the great moraines between the Swan Lakes basin and the Nass River gorge. At ten o'clock that night he located the Swan Lakes cabin. Someone had repaired the roof and left a generous stack of squaw wood near the fireplace.

Despite Rocky's warning, Greg built a fire and reconstituted a dehydrated beef stew. He ate recklessly, then slept until heavy hailfall awakened him the next morning. Examination of the cabin's interior in daylight made it plain that two men had occupied the dwelling for at least two months during the winter.

Torrential rain followed the hail. Greg pushed on north and let the storm take care of footprints. In the Fred Wright basin a noisy wolf pack inspected his camp, but identification was impossible in darkness and rain. Hoping for closer contact, he broke camp the next morning and followed spoor for ten miles to the basin's northernmost lake. Still not more than twenty miles from the Swan Lakes, he ate cold meals rather than risk campfire smoke.

With downright audacity that night, the wolves entered the general vicinity of the camp. They surrounded the area, sat confidently on their haunches, then shifted positions evasively, all the while whining and snarling.

As best he could and with slowest possible movement, Greg studied each majestic form through the binoculars at a distance of no more than eighteen feet. Inasmuch as he could learn nothing from so many black silhouettes, all essentially alike under the waning Moon-of-Fawns, he decided to chance a small fire to whet their interest and possibly inveigle them close enough to see if he could recognize definite wolves he had known at Nakinilerak Lake.

He broke out in a cold sweat, expecting Náhani to appear at any moment near the campfire. Long-legged shadows shuffled slowly back and forth and sneezed when smoke irritated their nostrils, but no wolf came closer. As a last resort Greg played the flashlight beam on startled adults and skittish yearlings. Females and their barely weaned cubs stampeded out of range.

No silvery-white wolf associated with that pack.

A grizzled old male, a basilic wolf weighing about 130 pounds, monarched the troop. He snarled, bared his fangs, and chomped continuously during the two minutes Greg used the flashlight. One terse command from their chief effected total silence. The pack simply dissolved into forest shadows.

Greg was keenly disappointed. The fact that his camp straddled the established runway of a specific pack proved decisively that Náhani would venture no closer than twenty miles east of the Fred Wright basin. She was too intelligent to court the disaster latent in trespass upon the territory of so formidable a clan.

When two days and nights dragged by without evidence of another pack in the vicinity, Greg climbed the long ridge north along the upper Kispiox basin. After crunching through twenty miles of friable shale in order to stay above timber line, he descended into the hazy-blue depths of the Nass River gorge – that enigmatic Kitiwanga quarter he had been unable to penetrate the summer before. On the shimmering stream's left bank he threw together a comfortable shelter in which to sit out a siege of rough weather. He studied his maps and wondered whether there still existed other regions on Earth so far from places with names. He knew, of course, that there were; but at that point he was worn out, hungry, cut, bruised, lonely, and possibly hunted. For a change of perspective, he took from the Trapper Nelson the one luxury he had granted space in the overweight pack: a copy of *Don Quixote*.

After the rainstorm he broke camp and plodded upstream, taking sometimes half a day to work out a ford across tributaries, waist-deep, wide, cold, and swift. When the cutbank river ran wall-to-wall, costly detours led either along precipices, through acres of devil's-club, around mile-long deposits of loose shale, or over massive wickerworks of driftwood.

Progress through each of these obstructions was not simply a matter of raw energy. There were pitfalls, any one of which could have ended the search for Náhani. Almost hourly he coped in a different manner with slippery, rotten logs, landslides, shaly cliffs, algal footing during fords of swift side creeks, fragile banks of névé, and unpredictable grizzly sows with cubs, or belligerent cow moose with calves.

Thanks to Youngpine's unforgettable schooling, Greg gave thoughtful attention to every footstep and situation. He employed the same care in plotting his daily position on the quadrants. Having relied so often on map and compass since college days, however, he had relinquished a measure of the hereditary Chimmesyan sense of direction.

But with all that exactness, he did not reach the Klappan-Nass headwaters unscathed. When he hobbled on to the beach of the lakelet that founts the Nass, he was suffering from a badly sprained ankle; lacerated legs, arms, and neck; and painfully swollen armpits from tick bites. Bruised and skinned, his knees and elbows were puffy from falls. Diary and logbook entries, written by campfire light under conditions of extreme fatigue and body pain at the end of each trying day, showed that his native exuberance and hearty enthusiasm were almost invulnerable to the physical assaults of the Kitiwanga.

'I must confess,' he wrote, 'I consider my hardest days the ones of greatest gain. There's always a song thrush in every cat's-claw brake. When I get to thinking the Plateau is hostile, a bear, fox, marten, or fisher comes along to share my rations or just sit for a friendly span. I have to laugh at the times I've knocked myself out over a tough spot only to find out afterwards there was an easier way through. I've got it soft compared with this Quixote fellow!'

*

Following a long rest at No-Name Lake and a slow trek over two Skeena Mountain passes, Greg once more undertook the rugged Stikine Plateau. He had neither seen nor heard a wolf after leaving the Fred Wright basin, although spoor and scats at 'wetting posts' indicated that seven major runways crossed the Nass. Track direction suggested that several organized packs had transmigrated to muskeg and tundra between the Skeena and Cassiar mountains in order to tagtail the northward movement of caribou and wapiti. The Nass river canyon, despite prosperous herds, was anything but typical wolf country. Getaway routes were too slow and uncertain for the liking of a wolf. Bears, coyotes, foxes, and wolverines now fished the early spawning runs; but wolves generally disliked raw salmon and probably found hunting more to their taste on highland barrows north and east of the Nass.

Wolf prints stippled the beach near No-Name Lake. Greg tinkered with the farfetched notion that he might *track* them into denning compounds, sleeping ledges, or runway narrows regardless of the terrain and enormous distances the packs habitually covered.

At the end of four days on a trail gone cold – without seeing or hearing one wolf – he suddenly stumbled upon fresh spoor. One extra large set of tracks had been made by a wolf much heavier than other pack members. And from the right front paw prints Greg immediately saw that the wolf had two missing toes!

The fresh tracks crossed a stream and mounted a sedgy, seepy hillside that eventually broadened into plateau savannah west of the Ross River – a mazy route Greg had roamed the summer before.

The wolves made no effort to conceal spoor as they did in regions where scent of trapper, steel, and gunpowder meant calamity. Because the latest tracks in the mud pointed downstream, Greg descended the north fork valley of the Ross, then tracked the runway to Cold Fish Lake where he awaited the pack's return.

He referred to his 'idle' days in terms of 'harvesting dandelions, cattail rhizomes, fern fiddle-necks, fungi, onions, parsnips, mustard, and seasonal berries to satisfy a craving for "greens".' Sorties into rocky upland braes for 'vegetables' made up for some of the time he admittedly 'frittered away on wildflowers'.

With July's Moon-of-Walking-Thunder in the wane, and no new signs of wolves, he reshouldered the Trapper Nelson and left the Stikine Plateau country in favour of the Pitman River valley where ten-day-old tracks of the two-toed wolf led. For sixty miles between Cold Fish Lake and the Pitman mouth, he walked without encountering serious obstacles.

There was no question about his having picked up a newly established timber-wolf runway.

8 Eyes in the shadows

Shortly before arriving at the Pitman–Stikine confluence, a narrows jam of driftwood debris completely spanned the river. Fresh spoor and distant wolf song increased Greg's optimism that he was now closing the distance between him and Náhani. In his eagerness he decided to take a chance and cross the river on the driftwood jam.

It was a risky business, as he well knew. In such situations, a cotter-pin log often held the entire framework together, and the slightest imbalance might shift that log and cause the jam to collapse. For that reason Greg generally avoided climbing over driftwood entanglements, and in this case the jam was higher than a full-grown cottonwood. He knew that he could not ford the Stikine and still survive the sucking current. Churning rapids and whirlpools below the jam discouraged the alternative of a raft. Therefore, it was now a question of getting over the log jam or tempting a more unpredictable devil farther downstream where the river would be wider and deeper. Tracks indicated that the wolves had made it over the jam.

The wickerwork held until Greg was within twenty feet of the right bank. Then a log rolled and dumped the young Indian into four feet of roily water. Fortunately the current at that depth allowed him to wade ashore with only minor damage to supplies. He was fortunate that only one log had been displaced. If the whole jam had given way, he would have been doomed. It was a sobering lesson.

The broad Pitman Valley soon narrowed and steepened towards an upland cul-de-sac where densely growing hardwoods, thorny vines, and craggy Omineca cols beset the route. However, a well-beaten game trail paralleled the left bank of the river to the lake of its source. The most recent tracks on the muddy trail had been made by the same wolf pack. Beyond Pitman Lake, Greg climbed a Pliocene glacial dyke six hundred feet high, a mile wide, and five miles long. That single rise separated the Pacific watershed from that of the Arctic. A longer lake on the east side of the dyke founted the Frog River, down whose right bank for three days Greg fought underbrush, wind-downed timber, and wall-to-wall water for forty miles.

One morning after breakfast he sat hunched over the map, trying to decide whether to strike south up the Kechika River or go on down to the Gataga. Both routes lay immediately west of the British Columbia Rockies. For some time there had been no wolf tracks. By now the wolves must have known that somebody was tailing them. Apparently they preferred to travel in the river, where no tracks would be visible.

Greg packed his rucksack and poured water on the campfire. When he turned to slip the map into a side pocket, his eye caught a flashing light from a granite outcrop to the northwest. Less than a quarter mile away, a cantering pack of at least twenty-five variously coloured wolves single-filed along the arête. He jumped to his feet and focused the binoculars.

An extraordinary silver-white female glided effortlessly through space and headed her troop towards the Gataga basin.

*

Two evenings later Greg reached Denetiah Lake. Could this at last be the friendly shore where Náhani had found some measure of freedom from pursuit? A well-defined wolf runway led to an extensive hillside forest where whelping dens faced the south beach an eighth of a mile above the lake's outlet. The closest burrows lay a hundred yards apart – mostly tunnel-like caves beneath stacks of sphagnum-entombed logs. Higher-level excavations under the extruded base of granite anticline were deep and well constructed. Others channelled through thirty to forty feet of soft sandy loam and invited disaster from cave-ins during heavy runoff. Many lairs served as auxilaries because no pack of twenty-five could possibly occupy all of them at the same time. Greg counted thirty-two caves.

There was no real clue that the queen wolf headquartered there. A distinctive manner of movement and the lack of visible sex organs – nothing more – suggested that the silver wolf Greg had seen might have been Náhani. A single black wolf with a slight limp had been near the head of the pack.

Spoor and scats were fresh – Greg estimated that the wolves had been there about six hours before his arrival.

After removing the Trapper Nelson, Greg hiked to a windy rimrock ridge high above timber line south of Denetiah. Here he scanned 250 square miles of forest and barren. Somewhere in that sprawling impassive hinterland, a silver-white wolf concealed her pack. The basin stretched into a blue-green wonderland of enduring distances and defeating bigness. From a rocky spur Greg explored through binoculars the misty chasms of the Kechika, Gataga, and Frog rivers.

He was about to return to the lake in order to prepare a camp site before dark, when smoke from seven small fires summoned his attention back to the labyrinth where the three rivers came together. Five minutes before, he had scanned that same section and there had been neither movement nor smoke. A warm, dry western downdraft gusted towards the Rockies with consistent force. The seven small fires soon joined and mushroomed into one Great-Council-Smoke, exactly as the Indians had planned it. A dry July Moon-of-Walking-Thunder, highly flammable resins of the season, and an oxygenating wind brought about rapid engulfment towards the ridges east of the Gataga. Given that the wind after sundown might change direction and move the holocaust towards Denetiah Lake, Greg had misgivings about setting up camp in the timbered trough. Embers could ride gusty updrafts across the labyrinth and ignite the Frog River forest.

For centuries Indians have ignited mature forest sections for access purposes, for improved game forage, and for introduction of new animal populations. Indian-set fires have been known to get out of hand occasionally in British Columbia; yet, the Indians are experts not only in gauging the forest acreage they wish to clear, but in prophesying changes in the weather. One could be reasonably certain that a downpour would occur within twenty-four hours after the fires had been set.

With that shaky assurance, Greg descended to the lake. The first early star shone as a tiny yellow spark over the wide sunset. There was something ominous in the air.

The raging fire twenty-five miles to the east, as well as the discovery near the Denetiah beach of finely crafted jasper arrowheads on the middens of archaic huntsmen, reminded Greg of Trapper-Dan Tall-Totem and his accomplices. With a twinge of worry, he sat by the campfire that night and wondered whether or not he had really escaped his pursuers – superior woodsmen in many ways when they weren't drinking. Their cause was far more fundamentally motivated than his, he conceded; there was a substantial money reward, invested time, damaged means of livelihood, and – for Dan's followers – deeply rooted superstitions.

With the Indian's knowledge of fire, Trapper-Dan might have encircled the vicinity with an offensive weapon from which neither Náhani nor Greg could possibly escape. The trapper's hatred by now must have reached feverish pitch. Whether he found the wolf or not, Greg determined during that sleepless night to find out definitely if the Tsimshian trappers were following him. The uncertainty and interminable waiting gnawed at him. Too often lately he had suffered the disquieting feeling that he was under surveillance.

Towards morning the rain came and extinguished the forest fire. Between two showers Greg climbed to the hilltop in order to assess the burn. A tremendous swath had been reduced to dark-grey mineral ash along the right bank of the Gataga River. He recalled that northern Indians sometimes set fires to facilitate placer prospecting.

From his rocky perch he laid out an itinerary to backtrack his route and find out once and for all whether he had been followed to Denetiah Lake.

More for rainy-day amusement than necessity, he constructed a driftwood lean-to with a sod and sphagnum roof. He topped the structure with flagstone in a way that no amount of run-off would leak through. He erected the shelter about fifty feet from the beach in order to take advantage of two well-spaced aspens for anchor trees. An adjacent cottonwood offered shade and held out a convenient arm from which he could 'string' his food supplies. After all that work, he never used the lean-to, even for sleeping, except during inclement weather.

*

When the Moon-of-Walking-Thunder promised several successive fair-weather days, Greg packed the rucksack and took off on a spying expedition of his own. The feeling that he was under constant scrutiny not only persisted, but grew. One entry in his diary at this time read: 'At times I feel deep physical pain, knowing that somewhere out there in the bush eyes are boring holes through me. A twig snaps, a pine cone falls, a trout leaps, a cougar calls his mate. Then absolute, ear-splitting silence grips the woods; every creature stops breathing. These little details were such a joy up to now. But *now* is a nightmare. I feel big-footed Sasquatch's icy fingers encircling my neck, tearing my limbs from their sockets. When I sleep I see old Trapper-Dan's blood-shot eyes and his boozy toad-eaters squinting down rifle barrels at me.'

So, he backtracked. The rain that erased earlier boot prints could reveal fresh ones. Using stalking tricks that Youngpine had taught him, he pretended to break camp and move on. A mile downstream from the lake the gorge narrowed. After minute examination of both canyon slopes through the binoculars, he removed every unnecessary item from the rucksack, wrapped the excess in his ground cloth, and strung the package from a bear-proof limb. If it was true that he was under observation, he wanted to be seen shouldering the Trapper Nelson.

Toting a much lighter load, he covered great distances, hid for hours at a time on vantage promontories, and scanned in scrupulous detail the route just travelled. At first he planned to hike the twenty-five miles into the new burn in hopes of checking out the Indians who had set the fires. But he reconsidered after weighing the possibility that even if they were still there – and weren't Trapper-Dan's men – they might speak neither French, English, nor Penutian–Athapascan. They might conclude he was a provincial agent about to charge them with forest firing, for which they could serve time in the penitentiary. Now and then such agents have met with quiet dispatch and deep burial.

So, he headed back towards the Pitman River watershed, haunted by the thought that Náhani had not contacted him at Denetiah Lake because Trapper-Dan with a loaded rifle had been hiding in the underbrush, anticipating such a contact. Greg's strategy was to execute wide, converging circles around the Denetiah trough. He avoided all sound and smoke, studied every game trail and open lea where a boot print would make an impression, probed every coppice where several men could hide a camp.

On the second day, in a platterlike valley surrounded by heavily forested hills five miles west of Denetiah Lake, he dropped quickly into a high-bush cranberry thicket. He sensed once again that he was being watched, but was unable to locate the source either with the naked eye or with binoculars. Unless he had unwittingly stumbled close to a camp site, he was confident of having exposed himself in no way. The forest duff, ordinarily as noisy as cornflakes underfoot, was still wet and spongy from recent rain. Slowly, crouching, he scanned the terrain, looking for the slightest movement, especially in shadowy or bulky masses of foliage capable of concealing a human body.

To relieve a cramped leg muscle, he was about to stand up when a twig snapped some twenty yards behind him. He spun around. Off came his pack. He whipped out his side knife and prepared for the first fight in his life.

Through a screen of brilliant rhododendron blossoms twelve inches above the ground, four pairs of yellowish-green eyes peered over black muzzles.

When the wolves realized they had been discovered, they backed silently into salal jungle and vanished.

Greg sat down and wondered. If sentinel wolves had kept him under surveillance at Nakinilerak Lake before returning to a denning complex, wouldn't they repeat the performance at Denetiah? He had been scanning the shadows for *human* figures, forms that would have stood between five and six feet tall. The wolves had eyed his movements from a flattened-out position hardly a foot above the ground, grey-brown coats blending perfectly with forest duff.

Still not fully convinced, he hiked to the game trailways at the bottom of the Frog River ravine, the only natural Kitiwanga migration corridor south of the Cassiars, north of the Ominecas, west of the Rockies, and east of the Stikine Plateau. That had been Greg's route to Denetiah Lake;

anyone else coming in would utilize the same narrow passage.

There were two trails on the Frog's left bank and three on the right. In the wet clay, sand, and loam of those five trails, every northern species on the move, except man, had imprinted his tracks.

9 The wolves come home

For a time after he returned to Denetiah Lake, Greg exposed himself to the scrutiny he felt but could not see. Slow-moving moose forged the sedge marshes, spiced their bland diet with spruce tips, then ignored the camp site as they waded into the lake. Elk and deer fought vicious duels for willow and alder shoots along the shoreline and paid no attention when Greg witnessed their battles from a distance of twenty feet. Golden eagles and great horned owls, about to compete for the same hare, utterly disregarded the young man's activities under their perches. Once convinced that the shadow men had not followed him to the lake – that no forest understorey veiled human eyes – Greg reverted to the serenity and trust of his childhood days, before he knew that men feared other men.

The summer sun as well as wind-sharpened light from the Moon-of-Walking-Thunder intensified the forest depths. The young man frequented the spongy aisles of spruce and hemlock, walking freely and without fear.

He had just returned with a harvest of fungi from below the lake's outlet on the afternoon of August 5. The mushrooms had been abundant, and Greg was so preoccupied with the feast to come that it took him some moments to recognize that a change had taken place around him. There were no animals along the lake shore or on the timber-line meadows. Birds exchanged new and strange notes. Jays, ravens, and waterfowl answered en masse when one ventured a signal. The succinct signals carried notes of alarm.

As soon as Greg sensed the change, he grabbed the binoculars and studied the hillside denning complex. No signs of life there.

But high on a broad, slanted shelf less than six feet below the skyline rimrock – 500 yards above the topmost lair – at least two dozen timber wolves lay asleep in the sunshine.

After observing and counting the motionless wolves, Greg went to the cook pit and kindled a small fire. Northern Indians do this when they anticipate a change. It is a ceremony akin to prayer. He saw no silver-white queen on the browrock, but his pulse beat wildly at the thought that this might be Náhani's band. Suppose it was? Could he expect, after three years, to meet the same gentle, patrician creature he had known at Nakinilerak Lake?

While he waited at Denetiah, Greg wondered why the wolves had stayed away so long from their own bailiwick. In all, their hunting range covered a maximum of 200 miles. They could easily have trotted this distance in four nights, hunted, feasted heavily, sunned themselves during the days, and returned, allowing ample rest for weaker subyearling members. Undoubtedly, the forest fire had driven them west. Greg's presence had been a disruptive factor, and the four scouts probably had not allowed the pack to return until they had completed their surveillance of the young Indian.

Apparently, when the scouts concluded that the man held no threat, the pack ventured back home. In logging the date, Greg rationalized that Náhani might have accompanied the scouts and partially recognized him. She might have led the entire troop to the dens with the assurance that this particular man smell recalled something vaguely familiar – something at least harmless, perhaps even pleasant. Greg was soon to discover, however, more subtle reasons why this particular wolf pack exercised caution to an extent he had not seen among other Kitiwanga wolves that had approached his camp sites.

The waxing Moon-of-Walking-Thunder had set over dim Cassiar crags. Four or five wolves sang briefly at midnight from a beach half a mile up the shoreline. Then total silence. A curious breeze nosed among the tall-shafted conifers. Greg stoked a small fire. Where was the nightly flute of the great horned owl, the cricket's fiddle, the tree toad's whistle? Campfire smoke infused the aroma of spruce pitch with wind-silvered driftwood. An atmosphere of expectancy seemed to magnify the solitude of that night. Greg leaned back against the log in front of the shelter. Softly he played the harmonica.

Appearing instantly, almost as if she had materialized from the fragrant campfire smoke, Náhani stood like a statue on the lake side of the log. When she uttered a deep-chested growl, Greg turned slowly and faced her. The campfire light shone in her wide-slanted, greenish-yellow eyes. Each motionless pupil was flecked with gold.

Only because she was there did Greg believe he had at last found her.

Cautiously, on all fours, he crawled around the end of the log. The huge white wolf growled deeply and chomped her jaws, but not savagely. When she raised her flews and snarled, Greg sat down and began to speak to her. Even from a distance of ten feet, she towered above him. Moving his eyes slightly, he saw the campfire light reflect from twelve pairs of hackled withers as her teammates surrounded the camp. Swiftly they faced him from well-drilled positions in the hunting encirclement. They stood and stared motionlessly, as if awaiting Náhani's orders.

'Náhani,' Greg said, knowing full well she could not recognize the name. But he hoped she would remember the tone of his voice. 'Come sit by the fire. Don't leave. I'm all alone. There's nothing to fear.'

She backed away until her hind legs were in the lake. Then she turned momentarily to lap several times at the water before pacing off into the forest. Her legion howled and chomped as they followed her. After several attempts at wolf song – and several answers – Greg threw his sleeping bag on the sandy beach. In order that the remaining sentinels, and perhaps Náhani herself, could see him asleep, he retired immediately. He tried to convince himself that somewhere in her subconscious warehouse there had been the faintest glimmer of recognition.

But he tossed for an hour, apprehensive that she had *not* recognized him – and never would. Her four-minute visit, like the first at Nakinilerak, could have been prompted by curiosity. Otherwise, why did she growl and chomp and run away?

He spent the next day on all fours in and around camp. Four sentinels, two on the hill north of the creek and two about fifty yards up the side of the denning hill, kept him under constant surveillance. At no time did they attempt to hide. For as long as Greg concealed himself inside the lean-to, one extremely large male left the hillside and paced back and forth along the beach within full view of the shelter's opening. When Greg emerged and sat on the log, the wolf returned to his post on the hillside.

At mid-afternoon, without ceremony, they changed the guard. Four replacements took up new positions. Throughout the day Greg had fed a small fire and played over and over again every tune he knew on the harmonica, yearning for some distant, primitive recall in the minds of the wolves. At Nakinilerak Lake they paid little or no attention to the instrument, but he believed they associated the sounds with him. He hoped they had heard no one play a harmonica since.

That evening half a dozen wolves trotted on to the beach about 100 yards west of Greg's camp. They strolled slowly towards the afterglow of an unusually spectacular sunset and yodelled half-minute songs. It was the first time Greg had observed wolves walking and singing at the same time. Ordinarily they haunched or lay on their bellies when they sang.

After dark the troop became somewhat bolder. Without

a sound, as if they were stalking an elk, they crept to within a few feet of the lean-to, then skittishly kicked up sand and leaves as they rushed away. For an hour and a half all was silent. A warm breeze hummed through the spruce needles and tiny waves clapped against wet sand along the shoreline.

At midnight Greg didn't try to hide or control a fit of shaking when he saw once again in the firelight the great silver wolf's eyes even before he could make out the full outline of her body. He decided to sit quietly and, if possible, send a friendship message between his eyes and hers, as they had at Nakinilerak. He felt tears of joy when she finally stepped around the other end of the log and haunched. She never took her gaze from his. He tried to count the toes on her right front paw, but this was impossible. He was tempted to use the flashlight, but decided against it.

Outside the campfire light he sensed more than saw a seething mass of bodies. An hour later, Náhani whined once. Her pack headed up the lake shore, apparently to hunt. Greg was thrilled that Náhani stayed behind.

When he placed half a dozen small sticks on the embers, the wolf rose slowly and walked to within three feet of his position – walked stiff-legged as a wolf does under circumstances of distrust. For a time she tested the scent his body emitted. From head and tail stance, from intense facial expression, and from a musical little whine in her throat, Greg surmised that she was endeavouring to associate his scent.

Whimpering almost imperceptibly, she padded cautiously about camp, sniffed, and pawed his equipment. She stood on her hind legs in order to sample the air beneath the limb from which the rucksack hung. Her action was his cue. The Trapper Nelson pack rack and canvas rucksack were the only items of his gear that had not been laundered since she had sprinkled them with urine at Nakinilerak. Slowly he crawled to the tie rope at the cottonwood bole. He lowered the pack. Náhani spent five minutes circling the sweat-laden canvas bag, aluminium frame, and webbed nylon straps. Although the pack bag and frame had been rained on many times – including several dunkings at rough river crossings – the fibres and metal frame still retained enough of her smell for recognition. Once satisfied, she opened her eyes wide, lifted her right hind leg, and urinated against the pack.

There had been no detectable fear as such in her attitude. From the beginning of the encounter, Náhani, backed up by her redoubtable phalanx, had the situation completely in command. They could have shredded Greg in less than a minute. At first there had been a definite aura of distrust. But almost instantly that distrust vanished when Náhani recognized her own smell on the Trapper Nelson.

Greg felt he was privileged to participate in one of the Earth Mother's great mysteries when Náhani approached him after she had sprinkled his gear. She looked him squarely in the eye and wagged her tail. Her eyes went soft and misty in the emberlight. Then she bounded mischievously out of camp and joined her hunting team somewhere up the beach.

To Gregory Tah-Kloma, that one moment made up for a winter fraught with insecurity and uncertainty, for two footsore and disappointing summers, and for an expenditure of more money than the reward offered for her skin. Now he could savour the knowledge that he had not only found Náhani, but also that she had recognized and remembered him.

Apparently the wolf led her pack to prey on that same night. They either made their own kill or robbed a cougar within forty miles, because they returned shortly after sunset the next day. From the beach runway that led to the denning hill, Náhani broke ranks and trotted immediately to Greg's camp. She haunched, whined, and wagged her tail within two feet of where he sat. He had ash-baked four large trout for her, but she nibbled only a few bites. The size of her belly indicated she had stodged possibly thirty pounds from a prey carcass. He set the fish on the log for other animals.

After bathing in the lake – a summertime habit following a feast – the wolves drank heavily, shook themselves, rolled in warm dry sand, and returned to the dens. Náhani licked Greg's hands, then joined her pack. Four sentinels assumed post and growled monotonously until he rolled out his sleeping bag on the beach.

*

Náhani and Greg had to get acquainted again – almost like human friends who have been apart for many years. She and her pack accepted his upright walking position, but with vocal reluctance. Yet it was soon plain that she experienced real joy each time they spent an evening together. Although she displayed more outward reserve than a dog, the big wolf demonstrated emotion in a similar way, with facial changes, ear and lip movements, catch-lights in the eyes, whines, head shakes, tail wags, pawings, and the compulsion to press physically close.

When she finally allowed Greg to run his fingers through her gleaming silver-white coat, he discovered five gaping scars that looked as though they had been caused by grazing bullets. But a grizzly's claws could inflict similar wounds. Two toes were missing from her right front paw, a graphic explanation of her reputed hatred for traps, if indeed she had been responsible for destroying Tall-Totem's winter lines. The distance between Denetiah Lake and Trapper-Dan's licensed route along the Skeena River, about 200 miles, seemed to discredit the trappers' claims – but not entirely.

Otherwise, Náhani was in prime physical condition. Her teats still indicated that she had never mothered a litter.

Innate strength, patience, courage, and a fierce will to survive, distilled from centuries of outrages against her species, accorded the wolf what it took to recover from grievous wounds dealt by man or beast, what it took to migrate hundreds of miles across an unfriendly Kitiwanga during the winters. Greg tried to dwell upon her pack's loyalty rather than the torture she had undergone – how they must have fought blood-scenting wolverines and other packs, how they fed her and licked her wounds, how they encouraged her and honoured her leadership across some of the most difficult terrain in North America. The disfigurement explained the pack's meticulously cautious return to Denetiah Lake.

Despite twelve-inch back and shoulder scars and the missing toes, Náhani did not limp. Her fluent leap and pace suggested that she probably had to slow down now and then in order for most members of her pack to keep pace. In a moment of exuberance, she once placed her hind paws on the top of Greg's naked feet and her forefeet on his shoulders, at which time he estimated her weight at approximately eleven stones, perhaps ten pounds heavier than she had been three years before. Most of her conclave weighed between eight and ten stones under a variety of coloured coats, predominantly grizzled grey. The remaining black wolf that limped weighed less. He still received warm attention from the other pack members.

A close study of eye and skin pigmentation led Greg to believe that Náhani's silver-white fur was not a form of albinism as he had once suspected but rather an atavism, a characteristic more remote than albinism, inherited from non-white parents.

Greg noted that young wolves in the pack expressed as much affection for Náhani as they did for their parents.

They generally ignored other adults. Possibly because a strong leader's authority and decisions are unchallengeable in a permanent *familistère,* her troop consistently mimicked her moods, even to stance and inflection of voice. There appeared to be instant, direct communication through eyes, ears, tail, paws, ruff, carriage, sounds, and emissions from scent glands.

Náhani delegated responsibility but never assigned authority. She ordered yearlings to guard whelps or the sick while adults were away cruise-hunting. Although the adults often filled the forest with sweetest melody and showed the warmest amiability towards one another, their society was not characterized by mercy according to human standards. The dominant hierarchy never permitted intramural fights. At Denetiah Lake, Greg saw two yearlings expelled from the pack because they challenged Náhani's authority by neglecting a den assignment while the main pack hunted. When the team returned and found the two youngsters romping on the beach, they attacked them severely and never allowed them to rejoin the troop. Náhani watched the proceedings with a faraway look in her eyes, but took no stand against the ostracism. The sirenlike wails of the yearling wolves hunted the forest until hunger finally drove them away.

Whenever the wolves prepared for the chase, they assembled and huddled on the beach near Greg's lean-to. They touched noses, wagged tails, vocalized, and sprinkled every upright object in the area. Those designated to stay near the dens assumed their positions before Náhani joined her team-mates on the beach. Each hunter knew his own prearranged duties. Teamwork, whether at chase or in defence, was their formula for survival.

When Náhani periodically rounded up the entire pack to cruise the full runway circuit, Greg believed she did so to train the young as well as to refortify territorial claims against pre-emption by neighbouring packs. At the assembly call on the beach, the wolves' amber-yellow eyes looked as if they could pierce prey. As Greg stood near the hunting muster, he wondered if, at any time in history, man-the-hunter's eye had glowed with wolf-the-hunter's inner fire. Upon return from a successful chase, these paradoxically shy, sensitive creatures wanted nothing but quiet.

10 The way of the northern wild

On occasion the pack returned with empty bellies. Then, if their leader seemed cross and ill at ease, the other wolves reflected her mood until they had rested for a day. At such times Náhani abandoned the regular runway and led poaching parties into the Gataga and Kechika rangelands of neighbouring packs. No wild-animal assemblage on Earth equals a hunting party of wolves for intelligence, stalking ability, basic senses, and endurance. Before leading her team onto territory defended by competing packs, Náhani invariably indicated which wolf would follow her, who would take the third and fourth positions, and so on, so that a chain of command was formed with responsibilities for defence, offence, and chase.

Her second in command was a hefty male who directed most routine matters such as assembly, sentry assignments, 'peck rights' adherence, and division of labour at the dens. He would assume chieftancy in case the monarch died or became incapacitated. Greg thought he recognized the big 'lieutenant' as one of those he had observed in the bushes before the pack resumed residence at Denetiah Lake.

The pack accepted Greg's presence because Náhani ordered it. The wolves seemed to view him as a minor annoyance, although he detected a degree of genuine friendliness in three or four individuals. When he sat with Náhani by the campfire, certain older adults wandered briefly in and out of camp or sat nearby on the beach. But the young at first were shy and wild, seldom allowing him to approach closer than forty feet. Unless Náhani accompanied him, yearlings took flight at full lope, yelping as if he had speared them. Then their parents generally came down from the hill, snarling in a threatening manner.

Greg eventually recognized wolves he had seen at Nakinilerak. In a *familistère*, physical characteristics are so similar that even an experienced observer might find it difficult to distinguish among a dozen or so wolves with virtually identical markings; but daily association soon revealed subtleties of difference in physique and character. Greg missed three older pairs he had formerly seen. Eight new adults, he assumed, were progeny of Náhani's troop, although several could have been admitted from other packs.

Náhani's temperament had undergone a certain mellowing. She no longer browbeat her female subjects as she sometimes did at Nakinilerak. Greg was privileged to study many close 'conferences' she held with her heavy-set 'lieutenant', whom she called on occasions to sit near the campfire log. At Nakinilerak, Greg had never known her to share in decision making.

Following these 'eyeball confrontations', the big subchief invariably assembled a small group of wolves in the marsh near the lake's outlet. After digging mice, voles, and lemmings from grass-lined runs for an hour – probably as a ceremonial gesture to deceive neighbouring spies who might be lurking in the north-shore forest – the squad disappeared down the canyon for varying lengths of time, up to three or four hours. During such absences, a messenger would sometimes return, and most of the pack would leave until dawn. Greg concluded that the original 'conference' concerned prey to be clawed and fanged away from a weaker neighbouring pack or to be poached if the prize was worth the risks involved. The wolves generally returned with full bellies, and, more often than not, one of the members would be wounded.

There was another change in Náhani's monarchy – she no longer 'barked' her orders. When she called a hunting assemblage, she stood proudly on the beach, head polled, tail wagging slowly above croup level, front and hind paws flexing. At once her team grouped around her, yelped, bounded, nuzzled, then remained quiet while she communicated through customary signals. Regardless of the mission she led – with subalterns, hunters, or with the entire pack for hunting or reconnaissance – the group followed single file behind her, each wolf's position determined by inflexible protocol.

Whenever she walked among her 'subjects', they went out of their way to lick her across the face, pet her with a paw, vocalize, and wag their tails. They still rolled over and offered their bellies, a sign of fealty and submission to her absolute authority.

One drowsy afternoon an old spindle-gutted doe elk stepped from the forest for a drink about a hundred yards west of camp. Náhani and most of the pack were asleep in the sun on a grassy esplanade at timber line above the dens. An alert sentry barked a signal. Within moments the wolf pack sprang through the timber and surrounded the unfortunate doe. Greg ran to the scene, but Náhani pounced in front of him to prevent his advance.

He stood on the beach and watched in helpless fascination while the queen wolf's chief 'deputy' demonstrated the extent to which the pack was organized once the team surrounded prey. The big wolf barked a command, lunged and seized the elk's muzzle for attention. Simultaneously four veteran hunters slashed low at the hind quarters for hamstringing. The old doe was so far beyond her prime she may have realized the hopelessness of her situation. She plunged towards the underbrush above the lake; but before she could take a dozen steps, she simply wilted as ten wolves attacked her back and neck. The kill from the time the first wolves lunged until the elk made her final payment for the privilege of having lived, took at most four minutes.

Within the next half hour, there was no snarling while

more than 400 pounds of meat disappeared. Two sentinel guards stood near the remains, but no wolf fed again for four days. During that time, the main body of the pack left the rimrock shelf only for water and excretion. Unaccompanied by her usual bodyguard, Náhani, however, descended the hillside each evening at dark for a fireside visit with Greg.

On the fourth day after the kill, Greg decided to see what the guard wolves would do if he pretended to take some meat from the elk carcass. He had covered less than half the distance between camp and the remains when he saw a large wolverine swimming quietly across the lake's narrow neck near the outlet.

The two sentinels at the carcass may have been a bit slow-witted that morning, or perhaps they believed they could handle the situation without calling for help. After spraying and thereby stunning both guards, the wolverine knocked down one wolf, then gave a sustained war cry not unlike a cougar scream as he headed for the carcass. Of course the cry and the volatile musk in the air brought the wildly excited pack within moments. The guard wolf that had been attacked by the wolverine staggered slowly from the scene. His chest and forelegs were dripping blood. Greg noticed a surprising lack of noise among the other wolves. Warily, they trotted in a circle around the carcass where the wolverine stood helping himself to a meal, but no wolf cared to be first to attack. Then through the binoculars Greg saw Náhani's vaulting leap, and the interloper was hurled to the sand. Recovering, the wolverine rushed through the encircling pack to the wounded wolf and delivered a cleaverlike slash to the victim's neck. The wolf died instantly.

Apparently expecting the death of Náhani's confederate to demoralize the kennel, the wolverine underestimated the queen wolf's leadership. She uttered one savage yelp.

The entire troop fell upon the invader. Within less than five minutes they had devoured the animal except for the shredded skin, skull, and musk glands. The pack proceeded to eat from the putrefying elk carcass as if nothing had happened – except for one female wolf that mourned.

After the wolves had feasted and returned to the sunning shelf, Greg walked towards the fight scene to observe the female who sat by her dead mate. She pointed her muzzle towards the sky and wailed pitifully until mid-afternoon. When Náhani came down at sunset, she accompanied Greg to the dead wolf whose body he dragged into the forest for burial under heavy cobble. The female ran up the beach, and he never saw her again. She may have left the region to become a loner, but more than likely she simply returned to the family den where she starved herself to death.

An interesting territorial situation developed in the early evening two days later after the wolverine incident. Náhani put in a nervous appearance at Greg's camp. At first she dozed fitfully. Then she paced back and forth, knocked over a pot of laundry water, and refused to sit down. At first he interpreted her behaviour as an outward expression of a communication problem rather than sensitivity to impending calamity.

As always he stopped what he was doing in order to give full attention to Náhani. Soon, younger wolves came to the log and whimpered, while every adult in the phalanx light-footed continuously between the denning hill and the beach. Their motion accumulated a dizzying rhythm. Náhani tensed.

Then Greg saw a neighbouring pack of wolves slowly advancing from the west end of the lake as if they planned to take over the elk carcass. Suddenly Náhani jumped over the log, issued a short squeal – like a shrill police whistle – and assembled her *familistère* on the beach.

Greg rushed towards the lake, but Náhani immediately headed him off. She reared, placed her two front pads on his shoulders, growled softly, and licked him across the face in a loving but stern warning to keep out of her affairs. She then returned to the assemblage near the lake. The entire pack was silent, motionless, awaiting Náhani's next move.

With firm resolution the silver-white queen wolf – six feet ahead of her formation of high-headed subalterns – cantered towards the doe elk's dull-red bones. The rest of the pack followed in a loosely deployed formation. The setting sun touched the distant Cassiars and sent reddish-copper rays over Denetiah Lake. The forested canyons were suffused with a smoky-blue haze. Although he shielded his eyes with the binoculars, Greg could see only black silhouettes – twenty back-lighted warriors advancing towards the elk carcass from the west.

It never occured to Greg to violate Náhani's orders. He stood on the camp log and waited the oncoming conflict. Forty feet west of the now fetid carrion, Náhani's command to attack brought a lusty war cry from every throat on both sides of the battle line.

The fight obviously was a decisive one. Not only could it end Náhani's supremacy among her own wolves but it might cost them their runway, their denning compound, even their confederation as a pack. Perhaps the anticipation of those consequences had caused her nervousness in Greg's camp before the confrontation.

One by one the invaders broke and retreated after the initial onslaught, leaving two of their number for burial. Náhani's wolves did not pursue. On the way back to the dens to lick their wounds, most of her noisy muster passed close enough to camp for Greg to observe their bleeding slashes. Only a few were unscathed. Náhani threw a single quick glance towards Greg as she herded her team on to the hill, where they remained and bayed until dawn. Their baying was not a beautiful sound like their songs, but it spoke of the way of the wild.

After breakfast the next morning, Greg walked up the beach to bury the two dead wolves, but nothing remained – not even the elk bones. Coyote, fox, fisher, and wolverine spoor indicated that he had not been the only spectator during the short but furious brawl. When he returned to camp, Náhani was sitting with her chief 'lieutenant' near the campfire log. Apparently a serious 'conference' was in progress. Greg took meticulous precautions never to disturb any relationships among the wolves. In order not to interrupt 'conference', he walked to the marsh beyond the lake's outlet.

Within half an hour Náhani joined him. The big grizzled male wolf climbed to the sunny rimrock where the wolf pack lay nursing injuries.

Mustering all his nerve, Greg decided to interfere in Náhani's affairs, and approached her to examine her for wounds. To his surprise she stood rigidly still during the probing, even when he touched a tender spot. She was bruised and swollen along the shoulders and hip, but he found no open cuts. When he knelt and put his arms around her neck and stroked her broad head and ruff, she licked him across the mouth. Accustomed as she was to most communication through the eyes, she looked for a long moment into his face.

'Let's go, Náhani!' he shouted. He knew it was a mistake to call her by name. The last thing he wanted was for her to recognize the word. Supposing a *hunter* were to call her some day and she were to respond, thinking that the man was Gregory Tah-Kloma.

They left the marsh, forded the creek, and climbed through the cool, dark cloud forest. Sphagnum moss blanketed ancient deadfalls and crept like dodder up the boles of living trees. This shrouded timberscape of fallen debris provided slippery footing for a man, but was an easy-pitched, deep-carpeted stairway for the wolf.

High on a gusty alpine terrace that overlooked hundreds of square miles dark with hazy timber, dewy canyons, and deep-set lakes, the two climbed. Greg never led the way when Náhani accompanied him on a hike. Either she led or they walked side by side. She selected a south-exposed ledge – a basement of granite – where they sat down out of the wind. The big white wolf shone with dazzling brilliance as she flattened out her body and placed her leathery black muzzle on outstretched paws. While he combed moulting guard hair from her coat, she snoozed with casual indifference.

Greg knew that Náhani was too regal ever to reveal that she hungered for affection.

11 The monarch of Denetiah

The late inter-pack warfare stirred general edginess among Náhani's 'lieutenants'. The bulky-necked males, whom Greg thought of as her 'cabinet officers', became skittish and quarrelsome. All the males carried crest and pastern hackles at full erection like lions' manes. The females swaggered night and day as if dispositions were already beyond the boiling point. The pack resembled a flock of mountain sheep about to reshuffle 'butting rights'.

Towards full moon, tension peaked. Instead of dozing away the sunny mornings on cisalpine browrock, Náhani and her 'staff' ran speedy patrols along the entire shoreline of Denetiah Lake, a running distance that approximated thirty-five miles with forest detours around impasses.

She interrupted campfire visits that week and left the log repeatedly for 'conferences' with her bodyguard in forest shadows. As excitement mounted throughout the pack, Greg was not surprised when the queen wolf came to him one night after sunset, reared, and placed her paws on his shoulders. She licked him across the face to 'tell' him that the pack would be gone until the moon waned. He assumed the wolves would carry out a thorough patrol along the full length and breadth of their runway. They would refresh all boundary monuments, re-establishing claim to their forage range against poachers. Greg knew that the runway could conceivably encompass 200 miles and could take them well within the range of Trapper-Dan.

Baying off steam-head excitement, the pack assembled quickly behind Náhani and headed up the beach. Greg could still hear the wolves when they left the lake shore near the southwest inlet. Their voices faded as they climbed and crossed the long ridge south. Like Alaskan husky teams, the wolves never changed pace regardless of terrain. By the time they reached the Frog River watershed, a sense of loneliness descended upon Greg. He played the harmonica and sat by the fire for several hours.

He had often wondered why Náhani had postponed full range cruise. Perhaps it was because the pack had brought down sufficient game near the whelping dens, which led him to believe that a wolf's territorial claim, whether or not the total area was necessary for hunting, might function as a symbol of status. Perhaps he was thinking in terms that were too human, but he felt quite certain that no wolf pack in the game-rich Kitiwanga actually needed a 200-mile range in order to eat regularly. He had seen bands race along outbound general runway trails – ignoring readily available natural meat – but stalk like Indians when they came within a mile of home base.

Greg had observed that Náhani's wolves habitually lived within a loose serfdom. 'Peck rights', property ownership, division of labour, and protocol remained rather static, but individual positions in the hierarchy were apt to change overnight. He remembered a statement made by a so-called authority: 'Wolves always travel a range circuit counterclockwise; they are prisoners of territorial "irons" like birds.' Unless Náhani and her company were extremely unusual wolves – which Greg denied – they often demonstrated flexibility. Adherence to habit could be broken instantly if a situation warranted it. In other words, a degree of sophisticated headwork quite often seemed to replace instinct. Greg further concluded that wolf teams, although composed of creatures who trusted instinct and distrusted innovation, conformed to very few fixed patterns outside the seasonal rhythms of biological functions.

Greg also believed that Náhani had some ability to reason. Apparently she had acquired her vast range because she knew she was a hunted wolf and therefore needed a wide variety of terrain in which her pack could hunt, gather berries and herbs, bathe, sun, play, breed, den, train cubs, and escape. When he had tracked the pack to Denetiah Lake, Greg had been able to follow the travelway with ease only when he could locate scent monuments in conjunction with the two-toed track. At intervals along the route – there being no path in the usual sense of the word – he was constantly on the lookout for trees, stumps, logs, bushes, stones, skeletons, and other objects that had received regular squirts of urine, droppings of faeces, wipes of paws, rubs of scent glands, and deposits of saliva. It was the wolves' habit of marking their established runway against intrusion by other wolf packs that led him finally to the Frog River sighting and the subsequent route to Denetiah Lake.

He was concerned that Trapper-Dan, who was reputed to be the most astute tracker in western Canada, would also have been able to decipher the runway signs that pointed to a denning complex.

*

The evening after Náhani and most of her pack left on the forage range tour, Greg received a pleasant surprise. Two older brindled females, left behind in the dens, half crawled into camp, whined, slavered, wagged their tails, and fell on their backs. These two had often made friendly gestures from the shadows, but Náhani never permitted them to enter the camp area. From external appearances, they looked exactly alike, closely resembling malamute German shepherd hybrids. They were considerably smaller than average, and probably weighed less than nine stones each. Still they were strong enough to divide a deer's spine with a single bite, or crack any long bone in a moose's body.

They were happy wolves. While Náhani was away, the two spent most of their time with Greg and became so friendly that he found it difficult not to treat the exuberant wigglers as dogs. He patted and rubbed them, scratched their necks and chins when they finally became accustomed to the human hand near their heads. He gave them food rewards only when they took it from his hands.

At the same time he feared Náhani might expel them from the pack should she return and find them in his company. Nevertheless, he also considered the possibility that the queen wolf may have delegated the visits, because both females raced up the hill nearly every time he gave them cooked fish. They refused it raw.

It occurred to him one day to follow them. Sure enough, they led straight to a nearby den where an old male was convalescing. He had received serious injuries during the interpack fight over the elk carcass. The two females were his wives, a rare occurrence of bigamy among wolves. Ostensibly, the old male suffered no injury to his appetite. Until Náhani returned, Greg caught and cooked a daily combination of trout, char, and grayling, which the females took to their mate.

He was fishing for the wolves on the seventh morning after Náhani's departure. Distant yelping and baying drifted throughout the basin long before the *familistère* single-filed from the southwest inlet and loped down the beach. Náhani led the pack by ten lengths. As usual, Greg met the wolves on all fours. It was the kind of greeting they could understand. The queen wolf stopped momentarily, wagged her tail, and licked her friend across the mouth, but jumped back and ran up the hill when he tried to put his arms around her neck.

Her first concern, he rejoiced to discover, was the wounded wolf. She showed her satisfaction with the two females by cuffing them playfully and licking their faces. But the two concealed their friendship for Greg when Náhani was in residence.

After careful inspection of their dens, the weary team

bathed briefly in the lake, then climbed to the sunning shelf with little more recognition of Greg than a chomp or a tail wag as they trotted past where he sat at the side of their path to the hill.

At dark a tired, but relaxed, Náhani slipped quietly behind the camp log. She licked the back of Greg's neck before he knew she was with him. The waning Sagamore-Council-Moon projected extra-sharp outlines that night. Forest creatures sang more softly when the wolves were around. For an hour Náhani lay on her back with her head across Greg's lap while he rubbed her belly. He combed out loose hair, gorged ticks, and matted burs. She allowed him to remove 'clods' of pine pitch from between her toes and foxtails from her ears.

Suddenly she jumped to her feet. With her nose pointed towards the beach, she indicated that Greg should follow her. He found it incredible that despite the mileage she had recently covered, she yearned for a moonlight walk. They had not walked together at night since Nakinilerak. He thought she may have acted from a dim, recurring flash of memory.

Then and later he became convinced that Náhani shared his fascination for exploring the northland by night. At night, the deep-woods lake and its immediate surroundings underwent the most subtle transformation. Changing shadow patterns, even the slightest degree of change in movement or light intensity, triggered Náhani's immediate interest. As a wolf she instinctively registered every change in her surroundings, but Greg also believed she was attuned to his own emotional nuances. At times she studied his face and wagged her tail or pressed close against his leg.

Innate sensitivity in both Indian and wolf predicated their friendship. Greg's experience with Náhani symbolized what Youngpine had taught him and stood for. When the wolf pressed against him he forgot the hardships he had faced in the Kitiwanga. He had grown to enjoy all the huntsman's primitive pleasures and hardships, without the huntsman's need to kill. It was enough to experience close contact with wilderness creatures on wilderness grounds and terms.

As Greg and the wolf walked leisurely through the woods, climbed a hill, raced madly down the beach, swam together in the lake, or sat in silence by the campfire, he also came to realize Náhani needed this non-wolf reprieve from her own responsibilities. Thus, a recognized need, one for the other, may have been the strongest link in their friendship.

*

Greg rarely knew when to expect Náhani. At times she appeared at sunrise and haunched for hours by his sleeping bag after having sat by the fire until greylight. Again, she trotted down the hill in mid-afternoon, at sunset, or at midnight. If she came after he had gone to sleep, she never awakened him. At times she insisted upon lengthy walks along the lake shore by moonlight but often preferred to sit by the campfire when there was no moon. By day she liked to climb the lofty ridges above the Frog or Turnagain watersheds.

On most occasions when they left camp, day or night, four to six shadowy outlines moved or stopped as they did. At first Greg resented the background bodyguard that never let the queen wolf out of sight. But then he thought of an ever-possible charge by a wolf-wounded moose, silvertip grizzly, wapiti, or caribou during one of their unguarded moments. Those silent, ghostly shadows – always at a discreet distance – became appreciated.

When Náhani elected to sleep among her own on the rimrock, Greg fished, hiked to the barrens for fresh 'greens', or explored the marsh. With store-bought staples dwindling towards the middle of August, he went on longer forays into berry brakes where he and a two-year-old black bear scowled at each other during competition for the fruit.

He wrote: 'The bear hates me because Náhani generally finds me when I come here. She chases him over the hill. I hate the bear because he browses the barrens day and night, leaving only the *sour* racemes for Náhani and me.'

From rotting logs and subsphagnum leaf mould, he collected daily about three pounds of morels, agarics, truffles, and puffballs to bake with trout and grayling. The bog yielded succulent burdock stems, purslane, horsetail tubers, onions, and fern fiddlenecks.

By experimenting with some of Youngpine's recipes, Greg created a weird bouillabaisse of fish, fungi, and fodder. However, he was never able to persuade Náhani to share his enthusiasm for it.

Greg had barely finished one of his 'outback table triumphs' one evening and had sat down on the log near Náhani. He was procrastinating about clean-up. A swarm of flying squirrels provided a sight and sound spectacle in the forest across the outlet. Suddenly Náhani growled, stood erect on the log, and hackled from poll to croup. Then she uttered a high-frequency, subvocal note, weakly detectable to the human ear. A sow grizzly and two cubs headed towards camp, attracted by cooking odours but apparently unaware of the occupants. Ten wolves materialized instantly, five at each end of the lean-to, as if they had been created on the spot.

Náhani, with raised flews, continued to issue subguttural warnings, some of which Greg could hear only if he watched the movements of her lips and jaws. The deployed phalanx stood motionless, erect, legs wide apart, ruffs distended, heads and tails in the battle posture – below loin level. When the bears waddled up within twenty feet, Náhani let out an explosive, deep-chested snarl. She followed the outburst with a bubbly whine when her ten braves stepped forward and crouched for attack. Their shoulder and hip muscles rippled, but they appeared more amused than alarmed. Greg felt that Náhani realized that only a wolf pack could change Ozilenka's course, a pack intelligent enough to bluff if possible. Ozilenka would bargain for the life of a cub.

As the bear reared and threw her forelegs wide apart in the bared-claw fighting stance, a halo of midges clouded her vision. She swatted the pests away from her eyes and calmly appeared to be assessing the number of wolves. After issuing a snortlike grunt – more like an exaggerated belch – she quietly ordered the cubs to trundle behind her. At a distance of fifty yards up the beach she turned, chomped her jaws, and trumpeted a threat that echoed throughout the basin and set off general alarms among smaller creatures.

When Greg turned again towards the lean-to the wolves had gone.

On the following evening, Greg and Náhani again sat on the log and watched the playful squirrels. An adult bull moose sloshed through the creek and displayed an obvious dislike for wolves the moment he saw Náhani. Shaking his rack, and with a terrifying bellow, he kicked up geysers of sand and threatened a head-on charge. Nothing equals a bull moose for noisy rage. Within moments of his appearance the wolves surrounded him in a running circle, nipping at his rear end. The moose set off for the forest, after which Náhani ordered an immediate withdrawal from the arena.

In the ensuing silence, Greg wondered why Náhani had ordered her pack away from a ton of red meat on her front doorstep. He concluded that judgement told her that it was too risky to engage in an encounter with a prime moose on open beach where the huge animal could step into three feet of water and kill every wolf that came within range of his sledge-hammer hoofs. She would never have allowed her pack to take on even an old, sick, or emaciated moose so near the whelping dens. Ozilenka would smell the blood and return to feast upon the carcass before the wolves could clean it up. There had been too much trouble earlier when her hunting squad neglected to herd the old elk a mile up the beach before making the kill.

In the wonderfully sane world of the wolf, adaptability and good judgement so far have saved the remaining packs from extinction; moreover, wherever the wolf has gone – driven before slings, spears, arrows, guns, traps, and poisons – prey species have not only thrived, they have improved. Wolves have weeded out those unfit to reproduce.

In Náhani's complex mind there may have been a further reason for calling off her hunters. After the dark of the Sagamore-Council-Moon, the several families at different times took their yearlings and subyearlings on their own expeditions, some of which lasted almost a month. Greg and Náhani had walked about five miles up the north-shore beach one night when a baying male, his mate, and three offspring overtook them. The farewell scene was touching. They all wagged their tails, licked Náhani over the face and brisket, whined at Greg's feet, then raced away into the night. Although the other families returned for autumn hunting rendezvous after their private cub-training trips away from the pack, this group apparently wished to establish a *familistère* of its own. The five headed for the Turnagain watershed, a landlocked Eden to the northwest. Perhaps they planned to tagtail caribou herds across Liard Plain muskeg and into the Yukon territory where the Selwyn Mountains give rise to the Nahanni River.

12 A song of the Kitiwanga

On nights when families or groups of two- and three-year-olds prepared to leave, they came to the lake and sang, perhaps half a dozen wolves in a circle. Their voices, of course, were not as deep as those of older wolves, but there was a purity of tone and key that was enchanting. The songs always began at the lowest notes, continuing in arpeggios to the highest. By the time the five-minute performance had ended, most of the other wolves on the hillside had joined in, pointing their noses skyward.

Greg had never heard Náhani sing. He would have recognized her had she sung, for her voice was pitched a full octave lower than that of the largest male in the 'honour guard'.

He observed that the wolves' various sounds and calls had special meanings for the pack members. Subvocal mutterings and squeaks were often uttered in frequencies too high or too low to be heard distinctly by the human ear. The wolves' extremely varied output – even to variations in tone for the same sounds – effectively indicated attack or withdrawal, affection or rejection, assembly or dispersal, agreement or disagreement, joy or sadness, consent or refusal, friendship or enmity. There were sounds to soothe, incite, and encourage. An entire vocabulary of murmurs governed the myriad gradations of gratitude, pain, pleasure, aggravation, fear, hate, jealousy, and suspicion. Signals to other species were probably confined to whines, snarls, growls, barks, and shrieks.

In addition to vocalizing the wolves chomped their teeth to produce a slap not unlike two paddles being hit – a stern warning. When they closed their lips and chomped, the sound produced was more muffled, like a stick being scraped along a picket fence – a greeting. Scratching the front or hind claws against surfaces like gravel, rocks, and tree trunks yielded a ratchet sound. This generally indicated impatience or other minor irritation.

Greg became adept at imitating the many snarls, whines, calls, and songs. With time and practice he was able to bring forth the desired response from the wolves when he issued a particular signal.

Body movements also conveyed the language of the wolves. The crouch, the flick of an ear or tongue, raised front lip, opened or closed flews, bend of the knee, tensed back, tail movement or position, and optical expression even to the width of eyelid opening were among the strongest signals. Body movements in conjunction with any of the several sounds communicated intensity of mood or intention to wolves as well as other animals. At times like these, the atmosphere was heavy with excitement.

Greg worried at first, fearing a general breakup of the pack. Perhaps he was to blame because Náhani spent so much time with him. At Nakinilerak there had been no such phenomenon. But in Náhani's expression he observed a calm and serenity from which he took reassurance. She made no effort to hold back any wolf that was preparing to leave the pack. In fact, for the first time her focus seemed to wander from her usual duties. Her team was a pack – a unit – yes; but first, last, and always they were mated pairs. And apparently, at least during the time that Greg observed them, the pairs were free to come and go as they pleased.

One night during the fading Sagamore-Council-Moon as Greg and Náhani sat by the campfire, he realized that he was now her only bodyguard. Not one wolf remained in the denning complex nor within calling distance. While the families were away, Náhani never left Greg's side. They ate, drank, sat, walked, excreted, and slept together. He and Náhani were alone. He shuddered to think what might happen if some misfortune befell the queen wolf during the pack's absence. The returning *familistère* would hunt him down and tear him to shreds.

Could Trapper-Dan have been waiting for this seasonal occurrence?

*

At dawn one day while several of the paired wolves and their offspring were still away, Náhani jumped over the camp log and ran, belly almost to the ground, straight for the beach. There she stopped and stood facing the west, with head and tail high, muzzle closed. She emitted a continual hum – deep throated, ventriloquial in timbre.

Greg craned his neck to peer up the beach without exposing himself to view, and saw a slowly approaching pair of yellowish-grey wolves. With infinite caution, the two strangers padded up to within ten feet of Náhani. All three wolves quickly raised their tails at right angle to the backbone and pumped musk from the caudal gland on the upper side of the tail near the base. This pleasant-scented musk, virtually undetectable to the human nose, is the wolf's most cordial greeting on the one hand and the most inviolate expression of identity on the other.

Once Náhani satisfied herself with the smell of the two wolves' salivary and excretory organs, she allowed them the same privilege. Then she snarled viciously at the female. Both strangers fell to their backs and exposed soft underbellies while she stiff-legged around and around the pair. Five minutes later she allowed them to follow her towards Greg's camp.

Nothing happened until the strangers got their first scent of the young Indian. Instinctively they crouched and froze. When he stepped into full view, the pair whirled and fled. In their haste to get away they seemed to reach full length with every leap. It was late that afternoon before Náhani coaxed them to the hillside.

Although Náhani appeared to invite the two wolves to join the pack, when the groups of hunting families returned she did nothing to restrain their hostility towards the newcomers. The unfortunate pair was relegated to the very bottom of the 'peck rights' scale. They were beaten savagely, repeatedly evicted from unoccupied dens, and the other wolves refused to share prey with them. At the same time, Náhani somehow communicated to the newcomers that Greg was an accepted member of the Denetiah Lake ensemble. While the attitude of the other wolves towards the young Indian was based more upon respect than trust, the new pair soon learned they could find refuge at his camp site during those first brutal days of initiation into the pack. When they were harried into camp of an evening while Náhani was sitting with Greg, she never drove them back towards the fierce jaws that encircled the camp and blocked escape. Eventually the pack accepted the newcomers and no longer harried them.

Pair by pair, most mated couples of the *familistère* had returned by the end of August. Each wolf resumed his place in the hierarchy. All yearlings and younger wolves returned with their parents, but several unattached individuals never came back. Presumably these found mates and joined neighbouring kennels or formed confederations of their own.

All the wolves were gaunt and knotty from late-summer moult, particularly the younger ones, who had undergone a strenuous apprenticeship with their elders. Training had begun, of course, as soon as the youngsters' teeth were ready. However, the young wolves had to be prepared to accompany Náhani and perform pack duties during rigorous winter hunting. Therefore, their elders had driven them to the brink of their endurance in late summer and early autumn.

They learned hunting techniques based on tracking and

exhausting the largest animals in the western hemisphere. Parents trained subyearlings to be vigilant, to take advantage of the camouflage of their coats when stalking, and to recognize every familiar as well as foreign forest scent. They masterered the message carried by tracks and paw prints. Attentive youngsters learned the art of killing and skinning a porcupine. Those parents who were the most dedicated teachers insured that their offspring would survive, should separation from the pack occur. Náhani would teach the teamwork required in running the open-end horseshoe formation for herding prey into ravines and other close quarters for quicker, more certain kills.

While other animals were preoccupied with storing up fat for the winter, the wolves feasted upon a bounteous berry crop. They dug rhizomes and bulbs, grazed late bunch grass, munched wood sorrel and cranberries, and gnawed sweetbark and bitterberry branches. From the first quarter of the Sagamore-Council-Moon until far into the Moon-of-Painted-Leaves they became vegetarians. On occasion ravens tried to lead the family teams to nearby prey for a cooperative feast during those annual meatless weeks prior to organized autumn hunting, but to no avail. In fact, during this vigorous yearling training, when the wolves selected deer and elk for attack, the animals were released at the end of the 'exercise' with only minor wounds. The wolves were densely furred non-hibernators, preying upon non-hibernators, and there was no reason for them to accumulate layers of fat against the winter months.

After the training period, all wolves resumed their regular routines. They took advantage of every possible sunny hour on the rimrock. During the dark of the moon, Greg and Náhani spent from eight to ten hours a day on alpine heaths not far from the pack.

With the waxing September Moon-of-Painted-Leaves, moaning chinooks herded sheeplike formations of clouds across the sky and soughed night and day throughout the conifer forest around the choppy lake. When a slow wedge of southing swans crossed the sky one afternoon, Greg moved into the lean-to and collected a supply of dry firewood. Time was running out. He fought desperately to hold back the clock.

The first general storm hammered against the Kitiwanga's enormous, malleable mass. For three days and nights the tempest raged; rain cascaded over the lake, lightning constantly whittled at sullen crests. Ancient spruce, long over-mature, crashed to the forest floor to become tomorrow's nourishing duff. Even during the most truculent squalls, the wolves rolled in mud, played up and down the beach, then bathed in the lake to drown their ectoparasites before returning to the dens. Náhani, dripping wet and smelling like new-cut wheat, came to Greg's lean-to day and night. She had begun to spend entire nights curled up alongside his sleeping bag while the wolf families had been away training the young. She continued the habit after their return. Somewhere, instinctively, she too must have known that time was running out.

On the morning after the storm, the north wind rolled the remaining cloud curtain back towards the Pacific. Most of the pack, led at full canter by the chief hunters, headed down the creek towards the Gataga River canyon. Náhani, on the other hand, indicated that Greg should escort her on an inspection trip along the north shore. A flight of shrikes, landing in two inches of water and taking to the air again at regular intervals, continually announced the position of man and wolf to the wild community. For some reason Náhani had sent her bodyguard with the pack. Since the wolves were invading territory defended by other packs, the guards might be sorely needed on the Gataga foray.

At a point where a fault line of upended limestone strata forced a detour into the forest, Náhani preceded Greg along the base of the extruded rock ledge all the way to the north rim of the Denetiah trough. At timber line they came upon an excavation, obviously the den of a fresh resident coyote family. Toothing bones, scats, and other fresh refuse lay scattered about the mound at the tunnel's opening.

Náhani stopped in her tracks. The crest hair on her neck and withers suddenly resembled a mane, and the long jowl hair stood straight out to make her face look twice as wide below the eyes. Rising and falling dewlaps and a backward flick of one ear signalled that Greg should remain motionless. He had never received that signal. Intuitively, he understood her meaning, having become sensitive to so many wolf signals. He watched her tense her hocks and pasterns. She spread her feet and barked, as only leader wolves do.

Nothing happened immediately. Then a drowsy male coyote walked slowly from the mouth of the tunnel. He took one look at Náhani and yawned extravagantly. Here was the first creature, Greg thought, to exhibit such a casual attitude towards Náhani. The two animals padded slowly in circles around each other, sniffing rear ends cautiously but with no apparent hostility. Obviously they were more than nodding acquaintances. Náhani stood forty inches high at the withers, was six feet two inches long from the tip of her nose to the base of her tail, and weighed more than eleven stones. She was almost twice the size of the coyote and four times his weight. Yet it was clear that the coyote neither owed nor paid homage to Náhani, the queen wolf.

When the coyote scented Greg, he drove deep into his tunnel. As Náhani continued on the inspection trip, Greg paused to compare her six-inch track with the two-inch coyote print.

Náhani quickened her gait along the steep hillside. Greg had to run to keep up. He could find no explanation for her sudden burst of energy, but sensed her urgency. Often she looked back to be sure he was still following.

Abruptly, Náhani stood alert. The main crest of the limestone fault disappeared under a quarter mile of grassy brae, a narrow belt of windy moorland between the lake basin's evergreen forest and the lofty alpine zone of frost-twisted subtundra. There was a gritty dryness to the air. In these austere barrows there were only two seasons: six weeks of spring and forty-six weeks of winter.

White ptarmigans, whose bronze-and-orange spotted feathers looked as if they were blotched with caribou lichen, stood on rocky outcroppings, thumping their wings in muffled unison. Fattening marmots whistled from sunny rockeries, obviously annoying the ptarmigans. A mother coyote appeared to be teaching her young to search out the silent, motionless chick ptarmigans. However, the pups were displaying greater interest in the plentiful supply of currants and *olallie*, the upland huckleberry. Then several yearling coyotes suddenly flashed across the heath and graphically demonstrated that they had learned their mother's techniques for stalking marmots and ptarmigan.

Náhani turned her head. She looked Greg squarely in the eye, a penetrating, thoughtful look as if to say, 'To a coyote your violets and avalanche lilies cannot compare with the smell of ptarmigan and margot!'

Greg put his arm around her neck. She licked the perspiration from his forehead and temples. More than ever Greg realized that understanding between him and the wolf was due in part to his Indian willingness to go along with her animal wildness, to the humility he always felt in her presence, and to an unexplored part of himself that was still primitive animal.

By now the sun was lowering behind the saw-toothed Cassiars. The air had chilled. Without disturbing the coyotes, Náhani indicated that Greg should follow her back down the hill. She had known all along where the coyotes were and what they were doing. Something deep within her animal mind had compelled her to take her friend to watch the spectacle.

*

Long before they reached camp that evening, Náhani conveyed a feeling that something was wrong. She kept looking back at Greg, urging greater speed down the north-shore beach. She barked, whined, or howled as she ran. She was clearly on edge. Upon smoothly packed cobble and sand the barefoot young Indian was scarcely able to keep up with the powerful wolf. Four large males ran to meet them when they were within a mile of the denning complex. Her 'staff-officers' reflected distress.

Náhani and her four 'lieutenants' plunged through the outlet flume. Then she inched carefully across the strip of beach, as though afraid and enraged at the same time.

The entire pack stood in trembling assembly at the log in front of Greg's camp.

13 A wolf betrayed

Náhani shouldered her way to the centre of the tightly compressed group. The four solemn dog wolves assumed guard positions outside the jostling corps. Greg waded the creek and slowly started across the fifty yards of sand and cobble between the outlet and his camp. He saw Náhani's long, white nose lift above the milling forms, heard her low moan that triggered high-pitched wails from every throat.

For a time Greg debated whether or not to go nearer. Ordinarily, clusters of the kennel moved quickly aside whenever he walked towards them. But now the picture had changed. As Greg stepped cautiously forward, the wolves' mournful cries dropped several octaves, and became deep, threatening snarls. In the rapidly gathering dusk, the circle of wolves swarmed and shifted so restlessly that he was unable to appraise the situation.

Suddenly Náhani's hackled white back moved away from the centre of the pack. Stepping slightly aside, she barked. The wolves slowly separated into groups of twos, threes, and fours. With bared, flashing fangs they howled and chomped, seemingly directing their rage towards Greg for something he did not yet understand.

Then he saw the fallen wolf. Panting on the white sand within a few feet of the camp log, lay one of the small mates of the bigamous male that had been injured in the fight over the elk carcass. Greg remembered the times he had fished and cooked for that family while the kennel patrolled the forage-range runway. He had deep affection for those two females, the only wolves in the pack besides Náhani that would allow him to touch them. Two inches above the pastern, the small female's right rear shank was firmly clamped between the jaws of a Number 4 double-spring steel trap.

Where had she stepped into that trap? And why with the *hind* foot? How had she managed to unstaple the trap from its anchoring clog and carry it back to Greg's camp?

The wolves, including Náhani, reached a peak of anger Greg had not believed possible in the so-called 'lower' animals. He had never looked upon wolves for what legend reputed them to be, but for what they meant to him. And now for the first time in his life, he knew – as an animal in danger would know – the meaning of fear.

Instinctively, and from bitter experience, the wolves associated the man smell with the steel smell of traps. Náhani, who at that moment stood at the head of their ranks and glared at Greg, carried a mutilated right front paw as a result of the combination of man and steel.

Greg realized that if he failed to face the wolves down, if he failed to communicate, his life would end in a matter of seconds. Without faltering, he walked rapidly towards the trapped wolf. He never once took his gaze from Náhani's icy stare.

The fleeting thought of the man smell – the *Indian* man smell – crossed his mind. Since the day when the Canadian Indian took up the white man's trapping tools, wolves had associated agony with the *Indian* man smell. Greg hesitated for a moment, sensing that the wolves were aware of his fear. And that very fear may have saved his life. His entire body exuded perspiration: chief source of the man smell they hated and shunned. Human faeces and urine somehow hold no terror for the wolf, but perspiration is another matter. Greg could only hope that Náhani would now accept *his* man smell.

*

'Náhani!' Greg shouted. 'I'm coming through!'

At the sound of his voice every wolf between him and the trapped female stepped aside and was instantly silent. But the victim quickly struggled to her feet. She lifted the trapped leg so that nothing but the three-foot chain dragged the ground as she hobbled away on three legs to her den. Her mate, the other female, and Náhani disappeared into the same tunnel. Not a wolf remained in sight. Greg stepped over the camp log, sat down, and cupped his hands over his face.

During those first moments after the confrontation, he was too upset to think clearly. He could think of no way to remove that trap before the wolves began to chew away the leg above the steel jaws. It would be no simple task. That female, while somewhat smaller than the average British Columbia timber wolf, could sever the human jugular vein and tear out the windpipe with a single snap of her jaws. Although insane with fear, shock, rage, pain, and fatigue, the wolf could still react with more deadly swiftness than any North American predator.

Deep in the cramped quarters of a den – possibly in the presence of several hostile wolves – the task of removing the trap would be many times more difficult than it would have been on the beach.

All through the long night he waited for Náhani. She did not come to his camp. Before retiring to the sleeping bag, he took the flashlight and went to the opening of the little female's den. Fifteen feet inside the entrance the tunnel made a right-angle turn, beyond which Greg was unable to see. The only sign of life from within was a faint, recurrent sobbing.

As he lay facing the stars, Greg tried to reconstruct the events of the day. The pack had left early, possibly to search lowland bogs for late-ripening berries, possibly to poach game along a weaker neighbour's runway. The wolves had travelled a minimum of fifteen miles by the time they reached the Gataga River. Even if the female had stepped into the trap at the point of confluence of Denetiah Creek and the Gataga, fifteen miles was a long way to hobble on three paws and carry the painful trap and chain on the fourth.

To each species, Greg thought, the Earth Mother has given some kind of armour to withstand her rigorous disciplines – but nature has never reckoned with traps and guns.

Greg awakened at sunrise. His urgency to remove the trap was spurred by Náhani's serious expression as she looked down at him from the log. There had been no serenades to greet the dawn. Greg rushed to her and examined her muzzle and brisket. No blood. She licked the stubble of whiskers that covered his chin.

He thanked Gitche Manito that she was still his friend.

The air was warm. He put on a pair of shorts, fixed a quick breakfast of corn-meal mush and dried milk, then rushed to the den. He held a flashlight in one hand, a pan of reconstituted milk in the other. Náhani reached the entrance before he did.

From the size of the dirt mound on the hillside below the tunnel entrance, he estimated that the whelping chamber lay about twenty feet beyond and slightly higher than the opening. The soil, a light sandy loam containing much glacial detritus and scratch-barked roots had caved in several times since the family had dug the den.

Greg listened. The faintest whine reached his ears each time he imitated a wolf sound for soothing or calling, but neither he nor Náhani could persuade the wolf to come out.

Greg always felt claustrophobic in dark, close places. Weather and insects permitting, he generally slept on the *outside* of his sleeping bag. Now he dreaded the thought of crawling into that crumbly tunnel. Yet under the circumstances, he had no choice. If he built a smoke fire, the wolf would simply rush out and enter another den. If the tunnel caved in, he would suffocate within four or five minutes.

He placed his flashlight and a pan of milk in the tunnel

beyond the entrance, and tried to squeeze through. His broad, naked shoulders were too wide. With a leather thong he tied his shoulder-length hair behind his head. With his side knife he enlarged the entrance and cleared away the dirt and rock. The tunnel was slightly larger in diameter than the entrance; but in order to inch along on his belly, he had to extend his arms full length ahead, pull with his hands, and push with his feet. Even so, when he took a deep breath of the rapidly fouling air, his chest completely filled the tunnel cavity. Apparently the wolves felt secure in such close quarters.

When he reached the turn in the tunnel, he flashed a beam of light into the whelping chamber at the far end. In a choking explosion of loose hair and dust, the male and the other female shot from the chamber into another tunnel with a secret opening somewhere in the hillside forest. Greg tried to back up but found it physically impossible to move backward. To make matters worse, Náhani had crawled in right behind him. Perspiration poured from his body as he struggled for oxygen, coughed, sneezed, and fought increasing claustrophobia. He was unable to bend either his knees or elbows. To add to the discomfort his hair had come untied.

Pulling himself into a cramped sitting position inside the whelping chamber, he handed the milk to the whimpering female. To his surprise, she gulped it down within seconds. Náhani squeezed in alongside her and sniffed the trap briefly. Not waiting for the dust to settle, Greg switched off the flashlight, managed to turn around, and re-entered the suffocating tunnel. At the bend he thought of taking the auxiliary route, but feared it might be even smaller. He could feel Náhani's hot breath on the soles of his feet as he pulled and pushed himself towards the circle of light and fresh air that seemed to be a mile away. Twenty-two wolves stood watching when he emerged from the tunnel.

Once outside, he and Náhani rushed to the lake for a five-minute swim. He suddenly became aware that the queen wolf was depending on him to do what she realized no wolf could do. He also had the feeling that she may have communicated this to the other wolves. At any rate, she appeared anxious for him to return to the den.

When he arrived, he was shaken to find that the little female had not emerged. If he could only persuade her to face away from him in one of those narrow tunnels, he could grab the trap, depress the springs with both hands, and free her without fear that she would turn and attack him.

If he was unable to force her to leave the chamber, he would have to bring in enough two-inch-diameter poles to fence her off while he reached underneath and released the rusty trap.

The other wolves continued to watch from a distance. They seemed to realize what Greg was trying to do. He made two more excruciating trips into the underground chamber, taking a pan of water each time. As long as he imitated certain wolf sounds, the little female allowed him to touch her head and scratch her chin; but if he moved a hand towards the trap, she snarled and chomped. Náhani also warned him.

The trap was old and rusty. It must have been forgotten by the trapper two winters ago. To open it might require more strength than Greg could muster. Trappers acquired the necessary leverage to depress the springs by 'breaking' the trap across a leg just behind the knee. There was no way of opening the jaws in the conventional manner inside that whelping chamber.

He suffered the same claustrophobia each time he squeezed through the long tunnel. On the third trip he foolishly decided to exit by way of the auxiliary passage. Halfway through he became entangled in a maze of rope-like tree roots. He was unable to fill his lungs with air. At the same time his leg muscles cramped. Finally he worked his hand to the sheath of his side knife by turning on one side. During the half hour it took to cut the obstructing roots away, Náhani lay at his feet and snored. Greg was too exhausted to enter the den again that day.

On the following morning a ruckus among the wolves awakened him. He rushed to the den. The little female sat at the entrance while nearly every male in the pack paced back and forth, growled, and sniffed the female's rear end. False oestrus, a condition sometimes brought on by the shock of a painful accident. Greg noticed that her trapped hind leg was badly swollen, a condition that probably eased some of the pain, but it also indicated that circulation below the pastern had been shut off. Gangrene would certainly follow, and within a matter of hours. When he approached with Náhani, the little wolf slunk back into the den.

Still dreading that narrow tunnel, and still emerging each time with his body covered with mud of his own sweat, Greg began to build a shield of short branch lengths to place between him and the wolf. She became more friendly with every visit until at last she did not snarl when he picked up the trap. But the presence of oestrual blood caused a constant tumult among the males. On every trip as Greg shoved the short poles ahead of him, he pushed aside one of several males occupying the tunnels. Their scrambling back and forth not only filled the passageways with dust and used up precious oxygen but also made the soft, sandy loam more hazardous.

Even before the shield between Greg and the female was finished, his flashlight batteries went dead. Total darkness as well as hot, heavy-breathing wolves in the tunnels now caused his claustrophobia to reach panic proportion. He probed under the shield for the trap. The springs were too strong for his hands alone. When he tried to get his knee under the trap, the wolf shrieked with pain, knocked the shield apart, and rushed into the auxiliary tunnel where noisy fights ensued. Although he should have known what to expect when he went to crawl out by the shorter regular tunnel, he bumped squarely into Náhani. An inch at a time he was able to push her and persuade her to back out.

Finally, in the open air again, he saw the little female dragging the trap towards the beach, at the same time fighting off her own mate and other males. Although mated for life, a male wolf will try to mount any female canid in heat – even a female dog or coyote – unless the oestrual period occurs among all females at the same time in a pack situation or separately during pack break-up.

With large rocks, Greg blocked both entrances to the den, then rushed to camp to get the nylon rope. The little female was rapidly communicating her distrust to the other wolves. Greg would have given up at this point, but Náhani, always at his side, seemed to bolster his hope. An hour later when the trapped wolf hobbled back to the den entrance and began to remove the blocking stones, Greg dropped a slip noose around the trap and threw the other end of the rope across an overhead spruce branch. Trusting Náhani to prevent an attack by the other wolves, he lifted the screaming female two feet above the ground by pulling the rope. He stepped quickly behind the tree, reached around with both arms in order to use the trunk for a fulcrum, grabbed the trap, compressed the springs, and freed the wolf in about seven seconds.

Greg ran immediately to the den entrances and cleared away the rocks, but the little female refused to enter. She ran in and out among the trees, shrieked, and slashed at every wolf within reach.

At last, exhausted, she seemed to calm. But there was neither delight nor relief in Náhani's expression. The panting little female moved from wolf to wolf, wagging her tail and whining. She was clearly out of her senses. Then she saw Greg with the roll of rope around one arm. Baring her teeth, she dropped her head and tail in the attack attitude and started slowly for the young Indian. He spoke softly to her as she advanced. Náhani stepped in front of him.

Ignoring the queen wolf's growl, the insane little female attempted to attack. With one mighty lunge, Náhani seized her throat and killed her.

While the other wolves looked on, the big silver-white stared into each face. Gradually the pack dispersed, some to the dens, some to the sunning shelf above timber line.

Greg picked up the lifeless body and the trap. Náhani followed. He buried the wolf in deep duff near the beach. Then he tied the trap to a low branch three-feet above the runway where the other wolves would have to smell it whenever they left the denning complex. He wanted every wolf to remember and hate that instrument for all time.

14 Reconciliation

For most of that day and night Náhani rarely left Greg's camp. Her mane and rump remained hackled. She snarled and chomped through open chaps every time a wolf passed within hearing distance. The other wolves reacted the same way. They were never still. Greg blamed himself for having jeopardized Náhani's sovereignty. For a week she was a despot, a threat to the very structure of the pack, slashing at her trusted 'lieutenants' as if they were intruders from a hostile band.

The air remained heavy with challenge. Every wolf's stride and gait indicated distrust, but those that dared the queen wolf regretted their choice. Nervous and taut in a confusion of instincts, the wolves seemed lost.

The wolf, of course, inherits his instincts; like a man, he also has the capacity for learned behaviour. For example, it is wolf instinct to avoid danger, yet curiousity often overcomes instinctive fear in many potentially dangerous situations. How well the wolf tempers curiosity with prudence determines the survival value of his learning, and thus the extent to which he has sharpened his instinct.

In that light, a mood of general depression struck the pack when they saw Náhani kill one of their cherished number. In their eyes the little female was justified in trying to punish her traditional enemy, man, for an unprovoked outrage against her. Their mood expressed itself through confused hostility towards Greg and Náhani. By nature wolves are intelligent, happy animals, expressive, open-faced, unashamed; but they are also finely keyed, deeply sensitive, and biologically responsible.

For a while Greg thought that Náhani was the most inflexible tyrant. She permitted no 'conferences' with the favoured 'staff officers'. She brutally beat every female who dared pass Greg's camp with even a suspicion of a hackled crest. She bobbed the tail of one yearling that answered her snarl in kind. An alliance of six usually docile males prevented her from killing the little female's mate.

But no wolf left the pack. On the eighth day after the unfortunate death, Greg noticed a gradual reversal of mood and attitude. Big males that had ignored his existence at Nakinilerak as well as at Denetiah came singly to the camp site and wagged their tails. They fell on their backs and whined while Náhani watched the young Indian prove to them that his hands were indeed pleasant when he scratched a brisket or rubbed flanks and belly. It was Greg's opinion that Náhani had communicated with her pack in a way yet totally unknown to man. Older females were even more demonstrative.

Shortly after sunset on that same day Náhani called for assembly near the outlet flume. The pack milled around her with some of the old adoration. She licked the chins of the dead wolf's sister and mate. There remained traces of uncertainty, confusion, and even hostility; yet her wolves were ready to follow her along the runway for the first autumn hunt and conquest of interlopers.

They had hardly crossed into the Frog River watershed when a gusting north wind seized the basin. Within moments the lake writhed under foamy whitecaps. The northlands most outspoken personality was the wind. Every creature bowed down sooner or later under its onslaught. When the wind blew, animal activities diminished; in fact, they ceased to exist. At 58 degrees 45 minutes north latitude and 127 degrees 30 minutes longitude, the Denetiah Lake region marked a transition zone between the vast muskeg voids of the Liard Plain to the north and the Rocky mountains backbone of the continent to the south. Transition zones were always areas of restive climatic conditions – colder of a summer, warmer of a winter, windier, rainier, foggier. Within the zone most life forms had reached a climatic climax of development beyond which the wind, directly or indirectly, intersected evolutionary processes. Still, a wolf family in the Denetiah watershed could live independently on 100 square miles of territory, while barrenland muskeg-and-tundra families required from three to four times as much.

On the heels of each atmospheric high-pressure system came the rain. While Greg slept one night, the first autumn blizzard struck. He worried because the wolves were now in September moult, unprepared to face frosty nights outside their dens. During the storm, Greg blocked off the open front of the lean-to, using his ground tarp. Following the icy rain, the wind returned. When the gusts subsided at sunset, he walked along the beach. Not a bird, squirrel, frog, or insect stirred. He recalled Youngpine's statement: 'There is loneliness in silence, even in a crowd.'

The next day he considered the idea of hiking out before the wolves returned. He studied the maps. He would drop down the Denetiah Creek canyon to the Kechika and Gataga rivers. Following the Gataga Valley towards the northernmost spur of the Rockies, he could take advantage of bare hillsides the Indians had fired the very day of his arrival. Beyond Ptarmigan Pass, a plain with no visible impasses led downhill through the valley of the Toad River to the Alaska Highway. A log trucker would give him a ride into Fort Nelson. A discreet telephone call to the garage in Hazelton would bring Rocky Longspear and the station wagon.

But Greg could not bring himself to leave. For several days following the storm, summer seemed to revive. Animal life returned. Waterfowl swam from the reeds and called with renewed vigour, so it seemed. The chorus of wildlife sang as if to make up for time lost during the blizzard. A whisky-jack that lived in a south-shore alder kept every creature advised on large animal movements. Late one afternoon the bird's strident whistle called Greg's attention to distant baying. With deceptively sleepy gait the wolf pack left the forest at the lake's southwest inlet and cantered down the beach.

Greg was elated to see Náhani not only in complete command but also the centre of loving attention when she stopped the pack at the camp site before continuing to the dens. He felt both surprise and pleasure that all members of the troop now eyed him and one another with only slight distrust. Actually, that was not quite correct. At some point along the hunting circuit, four strangers – two males, two females – had joined the pack as mates of unwedded three year olds. The new members were happy and playful, but at the same time suspicious and puzzled to find a man so close to their new homes. They reacted with seizures of hysteria and stampede, followed by awkward embarrassment when they encountered icy stares from the rest of the pack. The new members tried to entice their future mates as far from Greg's camp as possible when they played tag along the beach. Their courtships would not be consummated until late-winter oestrus.

In order to impress the newcomers with the extent of her dominion, Náhani resprinkled with urine the uprights of Greg's camp and most of his equipment. The new males and females rolled on their backs, whined, and wagged their tails every time Náhani appeared.

Three affiliates failed to return with the pack. Two unattached younger wolves had courted mates during that first autumn hunt, either to join other packs or to establish a *familistère* of their own. The limping old black male, mate of the wolf Trapper-Dan killed and skinned, had almost reached his maximum life expectancy of about twenty years. It was more likely that he died of a heart attack than a hunting accident.

Neither in conversations nor in his logs, did Greg intimate that wolves carried on exclusively in packs. Denning for parturition and brooding, he believed, was generally apart from pack activity – but not categorically so. Disciplined corps organization throughout the year is probably rare except in game-rich regions such as the Kitiwanga or under conditions of real threat. Records show that many leagues begin during late summer and last through the coöperative winter chase. Eight-week dissolution of the packs ordinarily occurs shortly before whelping.

The concentration of dens in compounds such as those at Nakinilerak and Denetiah lakes suggested that Náhani's pack, like others held together by unusually strong-willed leadership, operated as a unit the year around. Greg noted that Náhani denned regularly with her 'chief lieutenant' and his mate.

*

Because the Moon-of-Painted-Leaves brought clement days and nights, and because of increasing friendly encounters with the pack, Greg decided to stay at Denetiah Lake until he had completely exhausted his food supply or until such time as the wolves southed, if indeed they did. Certain minor intolerances still flared up now and then among the members of the troop, occasional ferocity and aggression; but the wolves – even the new pack members – now passed camp with friendly whines, closed flews, and eager tail wags. They no longer scattered like a covey of partridge when Greg walked among them. They sang their songs within fifty yards of camp.

But they stopped almost ritualistically and sniffed the suspended steel trap each time they passed underneath it.

The warm night of September 18 was one Greg would never forget. The moon was full. Náhani had slept most of the day on the sunning shelf with her troop. Greg had taken advantage of an especially greedy run of arctic grayling that schooled about the lake's outlet flume and rose for every lure that dappled the surface. The wolves had gorged themselves the night before on an early lemming migration, and for that reason Náhani ate only several small bites of a fish Greg offered her when she stepped around the log. She haunched beside him.

The stars shone with uncommon brilliance on this moonlit night, a phenomenon that often meant that the vagrant north wind was about to assert itself. After Greg had played the harmonica for about an hour, Náhani pulled at his jeans, her signal that she wanted to walk. As an alert woodsman, he sensed something extraordinary in her demeanour – a nuance of stance, one of those faint but perceptible shades of difference that he sometimes attributed to his own figments of fancy. She was a complicated being, and it often took patience to interpret her moods and intentions.

She led the way up the south shore, slowly at first. Instead of lingering behind in the shadows, her bodyguard of four magnificent males joined in the moonlight walk – two at Greg's side, two at Náhani's. How splendid they looked in their new fall coats! By the time the five wolves and the young Chimmesyan had walked a mile, one by one every wolf in the pack had emerged from a different aisle of the riparian forest to join in the walk. Twenty-eight superb denizens of the north woods were escorting Greg. There was some jostling for positions, even a few snarls and and snaps, probably because the new wolves were not completely oriented into 'peck rights' protocol; but at the end of two miles, Náhani, Greg, and the four 'staff officers' led the silent procession at a brisk pace for an additional five miles.

After Greg turned back, Náhani and her pack continued for an all-night inspection tour around the lake's circumference. He considered that particular walk with the wolves a supreme achievement, second only to the Náhani relationship itself. He gave her full credit for the pack's reconciliation after the little female's death. Gradually she had reconstructed trust that had been seriously shaken. In his diary he adamantly attributed nothing either to coincidence, caprice, or fancy.

He wrote: 'It was Náhani's superb control of the pack. She had the whole thing planned last night when she came into camp. Every wolf knew what was going to happen. Náhani now has shown me how a wolf thinks. In Youngpine's native Penutian legend there is an old saw that explains witch-doctors' magic: when anything strengthens a bond of friendship, Chimmesyans say the friends have walked in the shadow of a rainbow.'

*

Evening jaunts with the entire pack became a ritual to the extent that the wolves congregated each evening at the camp log and waited for Greg to join them. Day by day he sensed greater rapport with the *familistère*. He watched the complete disappearance of hostility within the group. At the dark of the moon, however, the pack left for a short hunting cruise and did not resume the evening walks until the light of the Fallen-Leaf-Moon. They had eaten well, so there followed a week of nothing but lazy hours.

One night a tom cougar dropped a spike buck on the north shore. The big cat ate heavily, then dragged the carcass into the underbrush beneath a giant cottonwood where he could guard his meals for the next week. As soon as the wind changed at greylight and blew from the direction of the bloody carcass, the astonishing chemistry of smell activated the wolves They began to bay at sun-up. Náhani commanded a selected dozen to cross the creek and proceed to the site. Through binoculars Greg kept abreast of the ensuing drama.

While four wolves kept the cougar treed, the others dragged the carcass to the beach. After eating their fill, they stood guarding the windfall until the other wolves on the south shore could answer Náhani's signal to feed. Suddenly she called retreat. She considered the cougar too dangerous to warrant the risk of allowing younger wolves to gnaw bones after the main body of the pack had returned to the south shore. The noisy, frisky troop followed Náhani into the lake and swam to their own bailiwick where they sprinted up and down the beach, barked, yelped, and whined until late afternoon. The cougar walked slowly to the remains of his kill. The big cat issued a scream that Greg heard above the din of baying wolves. After pacing his beach for an hour, daring the wolves to return – one by one – he climbed the forested hill and soon disappeared. For the next week Náhani did not allow any of her wolves to visit the north shore for fear of reprisals.

*

With shorter days and colder nights towards the end of September, plants as well as animals prepared for winter. Spruce cones, whose clusters Greg had barely noticed, were frosted over with sparkling resin as if they had been given a coat of varnish. Waxed-in crown buds glistened like golden scarabs from every treetop in the forest. Dwarfed cedars at timber line grew curious little wooden puzzles that ripened, cracked wide open, and dropped their seeds all in the same week. Red and flying squirrels chattered while they cut spruce cones and stashed them in leaf-mould middens. Upland rodents fattened on the season's abundance of rich seeds.

Responding to some inner signal, the monarch butterflies assembled along an inner-forest trunk. Another signal caused them to flicker like a thousand candles as they rose on a thermal into the exactly correct air current that would bear them to Florida.

The swallows that had swept the air so clean of insects assembled one morning and left for Honduras. Hermit thrushes, grosbeaks, and golden eagles left in pairs en route to California. Marsh hawks, buteos, and ospreys spiralled on Gataga canyon updrafts. Through his binoculars Greg watched them until they became tiny specks. They were on their way to Mexico's warm *lagunas*.

*

Greg observed a curious phenomenon. All through the summer he had been friends with certain robins, vireos, warblers, and siskins. He had put out food for them on top of the lean-to. For this reason they refused to go south with their flocks. For days they gorged themselves on pupated insects and seeds that had been blowing around since the Moon-of-Painted-Leaves. And now, after a week of feast and song, one by one each bird that refused to migrate dropped dead.

During September, hibernators and non-hibernators ate with only one purpose: to store up fat layers for the hunger-moons ahead. Greg had observed that the wolves ordinarily stalked and bedevilled neighbouring black bears in what seemed sadistic amusement. However, between September's full Moon-of-Painted-Leaves on the eighteenth and the Fallen-Leaf-Moon, full on October 18, the pack no longer molested the bears. They watched intently while the big clumsy creatures engaged in a peculiar ritual. Having chosen a single spruce trunk, they gnawed, scratched, rubbed, licked, patted, and embraced it, and urinated upon it. They woofed, bawled, moaned, and clucked during this mystic ceremony. While beleaguering the tree, they appeared to eat very little.

Only the wolves lazed through Indian summer. They knew that fattened prey would soon bog down in snow. Paired and tranquil, they seemed to enjoy their own company more during mid-autumn than at any other time since Greg had known them. Wolf serenades in the light of the Fallen-Leaf-Moon occurred with greater intensity and frequency than under earlier moons. Although pack members performed five-note songs with no predictable regularity, on clear nights Greg watched the wolves dogtrot to promontories where they threw back their heads and sang sixty-second solos, two-minute duets, and three-minute choruses with effervescent resonance and perfect harmony. Younger participants often broke pitch and yodelled. Between wolf songs, Greg played the harmonica for Náhani, but during this time he was never sure she enjoyed the music.

On the sharp, sunny midmorning of October 15, the idyllic autumn atmosphere was abruptly shattered.

Greg and Náhani were sitting face to face on the log, sharing ash-baked trout and grayling. Twelve wolves played tag and roughhoused on the sand between the log and the beach. The rest of the pack lay sprawled out on the rimrock, taking advantage of the last sunshine.

Suddenly a float plane stuttered across the deep river gorges to the southeast, swished over the treetops, and landed on the lake surface near the outlet. A gusting westerly had prevented any warning of the plane's approach.

At one cree from the whisky-jack, every wolf faded into dense forest on the denning hillside. There was no time for organized assembly. Greg saw Náhani lead the pack at full bay towards the Frog River watershed. It was too late for him to douse the fire and try to hide. The men had already seen him, the campfire smoke – and the wolves.

Terror struck the young Indian. The plane, exactly like the one at Tatlatui Lake, turned and taxied rapidly towards the beach in front of the camp site. Both hatches opened. Two men climbed out. One stood on each pontoon. With his left hand, each man clutched a strut; with his right he held a high-power, long-range rifle with telescopic sights.

Greg recognized both men.

15 The hunters

The bush pilot no sooner killed the engine than one man tied a rope to a strut and signalled to the young Chimmesyan. Obviously Trapper-Dan was not aboard, but Greg feared he was closing in from the bush to meet the airborne hunters. Why would any hunting party have selected the little-known Denetiah Lake?

Greg walked slowly to the beach. He caught the painter and pulled the aircraft ashore. Two middle-aged hunters with scoped Mausers jumped to the sand. The pilot followed. They stared at Greg in disbelief. The pilot was first to recover and offer his hand. Greg recognized him and both hunters as the three white men who had accompanied Trapper-Dan and Eugene Charley to Tatlatui Lake in the summer of 1966.

'Lex Morgan, Fort Saint John,' the pilot said. After an awkward silence he pointed to the hunters. 'Jay Spencer. G. Allen Ogilvy. Cheechakos from Seattle.'

In northland Chinookan jargon the term *cheechako* was used in a derisive way, meaning tenderfoot, white greenhorn, and worse. Apparently Spencer and Ogilvy were either ignorant of the inference or too absorbed in the size and number of wolf tracks on the sandy beach to be aware of their rudeness. They were also still recovering from the sight of an almost naked, long-haired Indian in the midst of a pack of timber wolves.

'They call me Gregory Tah-Kloma.' Greg spelled his name for them.

'If I hadn't seen you and that big white wolf sitting on that log,' Mogan said, 'I wouldn't have believed it.' There must've been a dozen wolves on the beach. At least that many on the hill. Náhani and the phantoms, Tah-Kloma?'

'I'd appreciate it, Mr Morgan,' Greg said, sidestepping

the question, 'if the gentlemen would put their rifles back in the plane.'

'We're here to *shoot* wolves, not let them get away – especially Náhani,' Ogilvy said. 'I want that skin. You can have the reward.'

'I'm not interested in any reward,' Greg said.

'Those wolves must be at least ten miles from here by now,' Morgan said. He was studying Greg with a mixture of curiosity and admiration.

Ogilvy was impatient. 'Let's get a move on. We've got shotguns in the plane and if we take off immediately, we can locate the wolf pack from the air and bank in at low altitude for a buckshot kill.' Noting Morgan's hesitation, he added, 'Don't forget, we're paying you to take us to Náhani. Are you with us or against us?'

'I'd like to talk to Tah-Kloma first,' Morgan said. 'Meanwhile, do as he says. Put the rifles back on the plane.'

The two hunters cursed but complied. They returned quickly to the beach and walked to the lean-to where Greg was getting into a shirt and jeans.

'What are you doing here at Denetiah, living with wolves?' Spencer asked.

'I'll answer all your questions in a minute,' Greg said. 'They may not be *satisfactory* answers. But they'll be honest ones. Shall we sit on the log by the fire? I'll make coffee provided you have the coffee. I take chinquapin tea. I pick it up there where the pack ...'

'What the hell's chinquapin tea?' Ogilvy asked.

'Indians drink it,' Morgan said. 'Native herb. Tastes like boiled flannel pyjamas. Trapper-Dan was always drinking it. Remember? I'll get the grub box.'

'Where did you come from, Tah-Kloma?' As he spoke, Ogilvy dilated his nostrils. Greg remembered that he had not bathed since the wolves had returned from the last hunt.

'I came from Kitwanga Lake. The answer to your other question is an Indian's secret.'

'Where's Kitwanga Lake?' Spencer asked. Instead of looking Greg in the eye, he squatted to examine the white shed-hair on Greg's sleeping bag.

'Southwest of the plateau,' Lex Morgan said as he arrived with the grub box. 'You're the guy the RCMP's yakking about. You're from Prince George. College man. You're the one all right. The wolf-man.'

'I'm a prospector,' Greg said.

Lex Morgan grinned as he prepared to make a pot of coffee.

'We've just had a spell of weather,' Greg said. 'I'll ask you to excuse the mess around here. I've been planning to clean it up.' He removed the pan of fish from the log. 'Náhani and I were having . . .'

A slip of the tongue.

'I told you that was Náhani sitting on that log with him!' Ogilvy shouted enthusiastically. 'Had to be. Biggest damned wolf in Kitiwanga. How many in the pack?'

'Twenty-eight,' Greg said.

'That's the killer pack. Killed that Carrier down around Hazelton last fall,' Spencer said.

'Impossible!' Morgan said. 'How long have you been here, Greg?'

'Most of three moons. There's no real proof Náhani southed out of the Kitiwanga in the last three years. This is the second summer I've lived with these wolves. Why didn't they kill *me* . . . if they killed Eugene Charley?'

'How'd you know about Charley?' Ogilvy asked.

'I could ask you the same question.' Greg said with a faint smile.

'This is what you call a real happenstance,' Morgan said. He laughed and shook his head. 'Spencer and Ogilvy here read about this "deadly" Náhani and the Indian that lost all his traps. A drunken bum called Trapper-Dan Tall-Totem. When these guys contacted Dan, he wouldn't go on the hunt unless Charley came along too. He said Charley knew more about Náhani than he did. Dan shook in his boots every time the word Náhani was mentioned. Spencer and Ogilvy footed the tab to bring the Indians to Tatlatui where the white wolf was supposed to have come from. They didn't care about the reward ... just the trophy. Both Indians swore they didn't know your name. Said you buddy-buddied with the wolves and would lead them to Náhani sooner or later. You were camped at the upper end of Tatlatui. Dan followed your tracks. Later on after you left, he found where you sat above our camp and watched us through the spyglasses. He was the maddest son of a bitch I ever saw. Too bad about old Dan.'

'What do you mean, "too bad"?' Greg asked.

'He and two or three Tsimshian trappers got crocked one night this summer. At some cabin near Swan Lakes. He shot and wounded a big silver-white wolf. The wolf rushed him and chewed hell out of him before one of the other Indians killed her. They skinned her and took the pelt to Smithers. Everybody swore it was Náhani. Tall-Totem put in claims for the reward but they didn't have enough proof. The wolf was not big enough for Náhani. I think it was about the middle of July when they hauled Tall-Totem to the hospital in Prince Rupert. He died a raving maniac ... rabies.'

After a long pause, Greg decided to tell the story. He told them about the white wolf he had seen leading a pack at Swan Lakes. He told them of the summer at Nakinilerak, of the long search through the Kitiwanga, of the winter at Kitwanga Lake. He spoke slowly and strung out his experiences with the wolves. Playing for time, he hoped that Náhani would be able to conceal her pack in the dense understorey of the Frog River forest. As Greg talked, the men sat quietly, listening with increasing fascination.

It was late afternoon when Greg finished the story. Ogilvy spoke first. There was a tone of respect in his voice. 'The last time we spoke to Trapper-Dan, he said "Náhani won't south again. If she leaves Thutade or Tatlatui, she'll go to Denetiah. If we follow her to Denetiah, she'll head for the Yukon. No man will ever kill Náhani." That's why we flew here. There were no signs of a two-toed wolf at Tatlatui or Thutade this summer.'

'There are lots of bush pilots, Greg,' Lex said. 'They've got wives and kids to feed. Plenty of trophy hunters will pay the fee to locate trophy skins. Bounty hunters believe Trapper-Dan killed Náhani.'

Spencer and Ogilvy exchanged glances, they seemed to have reached some decision. 'You can use this extra chow,' Spencer said to Greg. He smiled and emptied the grub box. 'We won't shoot your wolf. Náhani and her legend are already dead. The hunt is over. There's more pleasure in pursuit than in possession. We may even be able to keep your secret.'

'When do you want me to come in and fly you out, Greg?' Lex offered. 'You can't walk out after the end of this month.'

'Thanks, Lex. I know a way out. And thanks for the grub. I was getting low. I'll share it with my friend. I appreciate your change of plans. But this way we all win, including the big white wolf some people thought was Náhani.'

The hunters and the pilot returned to the plane. Lex Morgan circled the basin twice, dipped his starboard wing on the last pass over Greg's camp, then set his compass on Fort Saint John.

*

On the third night following the air-borne hunters' visit, Greg sat near the camp log. While he played the harmonica the first ghostly veils of the aurora borealis fluttered across the sky. Náhani walked slowly into camp and sniffed each new item of canned and packaged food. In order to allay her fears of the white-man smell, Greg opened a can of sliced peaches and shared them with her. As they ate from the same aluminium pan, her hackles lowered and she wagged her tail with customary enthusiasm for the things they did together.

'Náhani!' Greg shouted as he put his arm around the startled wolf's neck. 'He's dead! Trapper-Dan is dead! The Náhani myth died with him! We *did* walk in the shadow of a rainbow!'

16 Silence

On October 23, the dawn air had the chilly bite of tinsel rime. Steady breath from the Old Man of the North wheezed across the lake. Golden-armoured dragonflies manoeuvred through their last mating rituals, zigzagging over the lake. On the twenty-third, the big deciduous hardwoods dropped their colourful leaves to cover the drab forest aisles. Along the beach – even on the lake's choppy surface – the wind swept the leaves with a sound like fine sandpaper.

Greg and Náhani struggled up the gusty slope that had been his 'kitchen garden'. The hornlike peaks along the Rocky Mountain horizon were already iced with new snow. Even though her thick winter coat was now fully extended, Náhani bit Greg's hand and tugged at his jeans until he returned to camp and built a fire in the shelter.

On the twenty-fourth the wandering tattler, a talkative shore bird related to the sandpiper, announced the first complete blanketing of the lake with leaves. At the outlet flume mountainous leaf drifts accumulated. A noisy cloud of thousands of mallards swirled in from the Yukon tundra, descending upon the lake for a night's rest.

All creatures responded to the internal clock by which the Earth Mother times flocking, herding, prehibernation tapering of appetites, indifference to danger, and finally, migration. Squirrels became sassier and waterfowl reshuffled peck rights. Varying hares boldly confronted foxes with a belligerence they never displayed in other seasons, then bounded away. Birds and mammals stopped bathing.

*

Since the deaths of the birds that had neglected to migrate, muskrats along the stream had worried Greg with their procrastination at lodge construction. Then he discovered that they were waiting for black frost to cure their harvest of cattails and sedges. It was another example of the internal clock.

Each night, beginning on the twenty-fifth of October, frost brought its magical transformation. Dried stems became elegant banners whose white-selvaged outlines did not melt until noon. When evening frost tingled ears and noses, Greg and Náhani sat inside the shelter and leaned close to driftwood fire logs.

Anxiety permeated the air. Except for mountain goats, jittery herd of animals were on the move. Deer, sheep, elk, and caribou retreated into valley meadows far below their windy moorland summer range. Náhani waited until the herds had dispersed over lowland downs. Then she directed two expeditions along Denetiah Creek. After returning 'meat drunk' from both sorties, the troop spent a day chewing at tussocks of dry grama and Monarda on the heath below the rimrock. Diarrhoea throughout the pack indicated overfeasting on fawn and lamb.

On the morning of October 30, all wildlife tracks were rimmed with ice. Frost brought a new glisten to dun-coloured birch and aspen skins; frozen cattail, burdock, and eglantine stems at marshside swayed stiffly under a slapping breeze. And the frost did not melt. Greg awoke to a grey dawn with a diaphanous veil of ground fog that condensed, then froze, onto evergreen needles. All day the bouncy north wind played a brittle, tinkly tune among the fragile crystals. A single grebe that had postponed migration one day too long sat on the lake and called ceaselessly for its mate, who may have been caught by the great grey owl the night before.

At noon, on the last day of October, 1967, Náhani walked into the lean-to. The wind had stopped, but dark clouds seethed over the headland boulders. Greg and the wolf huddled together and listened to the whisper of the first flaky torrent of snow. The great grey owl with ice-powdered wings landed on the lean-to for a moment of orientation. As softly as the snow, he glided to the lake and silenced the remaining grebe's calls. Before sundown a determined south wind folded the clouds back towards the Yukon. The snow on the beach did not melt.

Greg dozed for a few minutes. He recalled a trance-like dream in which a shaman told him that his association with the wolves was not unique. Many Indians have known the same brotherhood under a variety of conditions. But his search for a specific wolf was probably singular.

A whine on the lake side of the log brought him suddenly back to reality. Náhani hadn't moved, but her entire *familistère* stood with heads and wagging tails lifted high as they faced the lean-to. In twenty-seven pairs of sparkling eyes the young Chimmesyan read the message. For two weeks he and the wolf had delayed that dreaded hour.

Proud and regal, Náhani allowed Greg to embrace her bulky neck. She pressed her head next to his. After licking his face and hands, she raised her muzzle and ululated softly. She leaped from the shelter, cleared the log in a single bound, and led her magnificent company across the flume and along the *north* shore. The wind unravelled their excited baying long before they disappeared over the Turnagain watershed ridge. As Trapper-Dan had predicted, Náhani headed for the Yukon.

Greg walked to the runway trail below the dens, cut the hanging steel trap loose, carried it back to the beach, and hurled it into deep water.

The last entry in his logbook read:

'Returned to the lean-to. Cut a skein of hair from my temple. Threw it into the fire as my ancestors the Ancient Ones once did. Thanked Gitche Manito for having granted me the privilege of knowing that wolf.'

Other Pan and Picador books that may interest you are listed on the following pages

Marie Herbert
The Snow People 75p

A fascinating year of Arctic adventure

Living in a remote settlement of Polar Eskimos, with her husband and baby daughter, Marie Herbert learned about survival – skinning a fox, making sealskin shoes and polar-bear trousers, and travelling behind a team of fierce huskies in howling blizzards . . .

The book contains eight pages of excellent photographs of life among the Eskimos, as well as a map of the area in which the Herberts stayed.

'She well conveys the atmosphere of blubber and blood, of huts steaming with wet skins as the wind rages' SUNDAY TIMES

in Picador Books

Margaret Craven
I Heard The Owl Call My Name 70p

Mark Brian, a young Anglican priest who has not long to live, is sent to the Indian village of Kingcome in the wilds of British Columbia.

While sharing the hunting and fishing, the festivals and funerals, the joys and sorrows of a once proud tribe, Mark learns enough of life to be ready to die.

On a cold winter evening when he hears the owl call his name, Mark understands what is to come . . .

An outstanding and much acclaimed first novel. The author's perception, wisdom and insight give her unique story the quality of a legend or fable.

'A rare clarity and simplicity. It is a long time since I was so moved by a story, touching in its dignity and wise in its folklore'
DAILY TELEGRAPH